T0137112

Cuatro Ciénegas Basin: An Endangered Hyperdiverse Oasis

Series Editors
Valeria Souza, Universidad Nacional Autónoma de México, Ecology Institute,
Mexico City, Distrito Federal, Mexico
Luis E. Eguiarte, Universidad Nacional Autónoma de México, Ecology Institute,
Mexico City, Distrito Federal, Mexico

This book series describes the diversity, ecology, evolution, anthropology, archeology and geology of an unusually diverse site in the desert that is paradoxically one of the most phosphorus-poor sites that we know of. The aim of each book is to promote critical thinking and not only explore the natural history, ecology, evolution and conservation of the oasis, but also consider various scenarios to unravel the mystery of why this site is the only one of its kind on the planet, how it evolved, and how it has survived for so long.

More information about this series at http://www.springer.com/series/15841

Valeria Souza • Antígona Segura • Jamie S. Foster
Editors

Astrobiology and Cuatro Ciénegas Basin as an Analog of Early Earth

 Springer

Editors
Valeria Souza (iD)
Instituto de Ecología
Universidad Nacional Autónoma de
México, UNAM
Mexico City, Mexico

Antígona Segura (iD)
Instituto de Ciencias Nucleares
Universidad Nacional Autónoma de
México, UNAM
Mexico City, Mexico

Jamie S. Foster
Space Life Sciences Lab
University of Florida
Merritt Island, FL, USA

ISSN 2523-7284   ISSN 2523-7292   (electronic)
Cuatro Ciénegas Basin: An Endangered Hyperdiverse Oasis
ISBN 978-3-030-46089-1   ISBN 978-3-030-46087-7   (eBook)
https://doi.org/10.1007/978-3-030-46087-7

This Springer imprint is published by the registered company Springer Nature Switzerland AG
The registered company address is: Gewerbestrasse 11, 6330 Cham, Switzerland

Preface

This is the fifth book of the Cuatro Ciénegas Basin (CBB): An Endangered Hyperdiverse Oasis series and targets the dynamic field of astrobiology. Cuatro Ciénegas Basin (CCB) is one of the most important astrobiological sites that exists on Earth, and in this book we examine how model systems, such as CCB, can serve as critical platforms to understand the origin, evolution, and distribution of life throughout the Universe. In this book, we begin with Chaps. 1 and 2 by explaining the field of astrobiology as well as its development in Mexico. However, as we only know of one experiment of life, planet Earth, locations such as CCB become critical to understand what makes a world habitable. Chapter 3 examines the origins of life and what were the probable conditions where the Last Unique Common Ancestor, or LUCA, arose. Following, Chap. 4 explains why stromatolites serve as important reservoirs for the study of life's biosignatures, and Chap. 5 shows the geographical distribution of stromatolites in modern times. However, to understand the formation of stromatolites and their microbial diversity, Chap. 6 examines the rare biosphere and discusses the hypothesis that in the beginning of diversification, all life was rare. In Chap. 7, the authors examine bacterial communities living in hot and high-pressure environments in hydrothermal systems at the ocean floor, which may harbor relict physiological capabilities that show microbial life in its extremes and also may reflect conditions similar to Jupiter and Saturn moons. Meanwhile, Chaps. 8 and 9 show stromatolites in different sites of Latin America, including the Andes and additional sites in México, with the aim of comparing these important astrobiology-relevant ecosystems. Chapter 10 presents the Mexican contribution to understand the origins of life and their current lines of research that look at the first stages of the evolution of life. Finally, the closing chapter is about the Archaea at a particular CCB site and discusses why CCB is such a special site for Astrobiology.

Although CCB is not as extreme of an environment, compared to Mars or the moons of Jupiter and Saturn, it is a special place for astrobiology due to its long and ancient history. CCB is a lost world that has survived and retained the ancient lineages inside the deep aquifer in the Sierra under conditions akin to an ancient, magmatic influenced, ocean where phosphorous was extremely limited, sulfur was rich, and oxygen was poor. In the case of CCB, under the mountain magma heats the

v

deep aquifer and water, along with deep microbes and minerals, surges to the surface, forming ponds, lakes, rivers and pozas. This unique oasis creates an ecosystem where stromatolites and microbial mats flourish under the sun. The CCB, therefore, represents a co-evolving community where the feedbacks between microbial life and the environment drive the habitability of this unique ecosystem. The underlying mechanisms that makes CCB an analogue of early Earth and Mars is precisely what makes life so diverse, the extreme low abundance of phosphorus, a rare element in the first three billion years of life. Despite the skew in stoichiometry balance, or maybe because of it, CCB is a hotspot of biodiversity. This amazingly diverse ecosystem has been studied by a large group of scientists and has been the theme of a series of seven books published by Springer.

We believe that CCB is an ideal analogue of Mars, a place where the ancient shallow ocean probably was poor in phosphorus and rich in the same prebiotic cosmic soup that made the origin of life possible on Earth. It is also possible that if Mars life evolved, it may also exhibit chemolithotroph metabolisms, paralleling that of modern sulfate-reducing bacteria and methanogenic archaea. Although the recent missions to Mars have not been tasked to specifically look for life, these missions have been quite successful in evaluating the potential habitability of Mars. Understanding what makes a world habitable is the extraordinary gift that the CCB deep aquifer can show us and help inform what characteristics or biosignatures to look for on other worlds. Nevertheless, CCB is succumbing to overexploitation of its most precious resource, water. Therefore, the ancient life in this system is slipping away in front of us. The overall message of this book is that we can stop degradation of our planet and protect critical habitats that enable us to learn about the boundaries of life on this planet and beyond. Perhaps, we can save humanity from destruction if we dare to imagine another society, a society that abides to life and tries to understand its rules and follow its lessons. In the case of CCB, we as scientists have been working towards improving awareness of this precious ecosystem through art, education, and science. It is our overarching goal to improve the public's understanding, appreciation, and love for the unique life, beauty of the oasis at Cuatro Ciénegas Basin, as well as increase awareness that such unique ecosystems needs to be nurtured and protected.

Mexico City, Mexico   Valeria Souza
Mexico City, Mexico   Antígona Segura
Merrit Island, FL, USA   Jamie S. Foster

Contents

Chapter 1
What Is Astrobiology?

**Antígona Segura, Sandra Ignacia Ramírez Jiménez,
and Irma Lozada-Chávez**

Abstract Astrobiology is an inherently multidisciplinary field that is focused on the origins, evolution, and distribution of life throughout the Universe. The question of whether life extends beyond Earth was a question that used to be answered mostly based on human imagination reflecting our passions and fears. Philosophers, scientists, and even politicians, such as Winston Churchill, have argued about the existence (or nonexistence) of alien life in the Universe. For scientists, this ambitious endeavor begins with Earth, as it represents the only known example of life in the Universe. Understanding Earth is, therefore, the first step to understanding the requirements for life to emerge and make a habitable world. In this book, with the collaboration of scientists from many disciplines, we gather the knowledge about the requirements, diversification, and characteristics of terrestrial life, as well as the characteristics of potentially habitable worlds in our Solar System and beyond. In this chapter, we describe the objectives and strategies of this dynamic field that has emerged with a multidisciplinary approach, leading us to one of the most exciting goals: the search for extraterrestrial life.

1.1 Aliens Everywhere

In Mexico, 41% of the population thinks that unidentified flying objects (UFOs) are evidence of extraterrestrial civilizations (INEGI 2015), while in Germany, the United States, and Britain, more than a half of the population believes extraterrestrial

A. Segura (✉)
Instituto de Ciencias Nucleares, Universidad Nacional Autónoma de México, UNAM, Mexico City, Mexico
e-mail: antigona@nucleares.unam.mx

S. I. Ramírez Jiménez
Centro de Investigaciones Químicas, Universidad Autónoma del Estado de Morelos, Cuernavaca, Morelos, Mexico
e-mail: ramirez_sandra@uaem.mx

I. Lozada-Chávez
Interdisciplinary Center for Bioinformatics, University of Leipzig, Leipzig, Germany
e-mail: ilozada@bioinf.uni-leipzig.de

© Springer Nature Switzerland AG 2020
V. Souza et al. (eds.), *Astrobiology and Cuatro Ciénegas Basin as an Analog of Early Earth*, Cuatro Ciénegas Basin: An Endangered Hyperdiverse Oasis, https://doi.org/10.1007/978-3-030-46087-7_1

life exists (YouGov 2015). Every year, movies, books, and TV shows display extra-terrestrial beings, some of them are heroes, others are villains, and others just live their lives until we find them. Aliens have all kinds of shapes, colors, and sizes. In all cases, what we learn from these mediatic products is not about life elsewhere but about our humanity in the face of "otherness." Our mythology about aliens often includes wrong ideas about scientists and their search for extraterrestrial life. The image of the all-knowing scientists that fight against the establishment to convince the rest of the scientific community that extraterrestrials exist is not accurate. The scientific community has created a science to understand life on Earth and to better define how to find it elsewhere. This science is called *astrobiology*. NASA's defini-tion of astrobiology is the study of the origins, evolution, distribution, and future of life in the Universe (Hubbart 2015; NASA 2018). Astrobiology articulates the knowledge about origins, the characteristics and evolution of life, Earth's history (its geology and chemistry), planetary bodies in our Solar System, and planets around other stars to understand life as a universal phenomenon. For the first time in human history, we have the expertise and technological tools to test the hypoth-esis: the origin of life on Earth is the result of a series of natural events; therefore, it is possible that there is life outside Earth. This chapter describes the main questions, target places, and strategies used by astrobiology in its quest.

1.2   What Astrobiologists Are Looking for?

The main problem in searching for extraterrestrial life is not the "extraterrestrial" part but "life" itself, as scientists have no consensus for a definition of life, despite of having hundreds of proposals (e.g., Trifonov 2011). What we have in practice is a list of characteristics, for example, reproduction, information, and metabolism, but we do not know if they should be considered valid everywhere in the Universe, as it is the case of the laws of physics or chemistry. A definition for life should set a clear boundary between the objects that belong to a "natural category" and those that do not, and at the same time, it should not leave anything that belongs to that "natural category" outside those boundaries (Tsokolov 2009; Mariscal and Doolittle 2018). The problem of defining life is twofold, since it has epistemological and scientific considerations, which sometimes have led to suggest that such endeavor is almost impossible, useless, and even counterproductive for the daily work in life sciences (Cleland 2012; Mix 2015). Epistemologically, our attempts to define life fail when (Tsokolov 2009) (a) we use undefined (or highly contextualized) terms, e.g., the term "information" has different meanings in biology and cognitive and computational sciences, (b) confuse a description or a list of properties with a defi-nition, or (c) define life arbitrarily in terms of minimal living systems (such as pro-tocells and viruses). Scientifically, we have only one sample of life in the Universe so far, the so called N=1 problem, so that our current attempts to define life are based, and thus biased, on the features, processes, and patterns of "life as we know it" on Earth. Also, life on a planet or a moon can only exist as a "historical-collective phenomenon"; we should thus conceptualize life on Earth as a single individual, a "life-individual" belonging to a universal collection (Hermida 2016; Mariscal and Doolittle 2018).

To tackle the previous problems, a universal definition of life should embody a Theory of Life, such as the definition of water – a substance that we recognize as a transparent and odorless liquid – uses the valence bond and molecular orbital theories to understand its basic properties. Both theoretical frameworks allow us to unambiguously define water as a molecule made of one oxygen atom and two hydrogen atoms (H_2O), so that we can identify it throughout the Universe regardless its physical state as liquid, solid (ice), or gas. During the last 20 years, philosophers and scientists from diverse fields (particularly astrobiologists) have been working on the construction of a Theory of Life, a major unifying explanatory framework of life, which is commonly referred in the scientific literature as "universal biology" (Hermida 2016; Mariscal and Fleming 2018). The program of universal biology is not just about finding out which features, processes, and patterns of life on Earth are universal and which are peculiar to it (Hermida 2016). Among other things, universal biology should also help us to clearly distinguish the different concepts within the life definition, so we can guide our understanding of and search for non-terran life. For instance, most of us are familiar with the concept of organism or cell, or what we understand for "being alive"; however, "being alive", "evidence of life" (e.g., a biosignature), and "life-individual" (e.g., life on Earth) are only incomplete representations of the concept of life as a phenomenon elsewhere (Benner 2010), regardless if such phenomenon is uniquely found on Earth or is widespread in the Universe.

The discovery of at least one other form of alien life would, without doubt, provide significant advances on the program of universal biology and the creation of a consensus for a definition of life; but until that happens (if so), cleverly our best option has been the use of "operational" definitions of life. They are based on the known characteristics of living systems and serve to build strategies for the search of life on places different from Earth (Tsokolov 2009; Benner 2010).

Our working definition is based on the most general characteristics of life on Earth. It seems that life on Earth covers a huge variety of forms, colors, sizes, or environments that could represent a challenge to classify but looking closer to some general patterns emerge. All terrestrial living beings undergo Darwinian evolution since their informational molecules (DNA and RNA) are diverse due to mutation, which means that everything that has ever lived on Earth has somehow common ancestors that used these molecules (see Chap. 3, 6, and 10 in this volume). Looking even closer, the chemical elements used to build useful molecules for life are just a few; they are famously known as CHONPS for carbon, hydrogen, oxygen, nitrogen, phosphorus, and sulfur. Carbon is the atom in charge of putting together the rest of the elements because it has the property of bonding with up to four elements including itself, by a variety of chemical bonds. In this way, carbon serves as the central chemical atom for the matter that makes living beings. Molecules also need to move to find other molecules and react to finally build living beings. The medium they use is liquid water. This is true for all life forms on Earth. Thus, we can say that life on

Earth has a chemistry based on carbon and requires liquid water. When astrobiologists say "we are looking for life as we know it," we are not thinking on eyes, antennas, or gray humanoids, but in carbon-based molecules and liquid water.

But people may think maybe there are other forms of life that we have not imagined yet. First, carbon is the fourth most abundant element in the Universe, and water is composed of the most abundant element, hydrogen, and the third most abundant element, oxygen. So, life is not made of anything special. For the second reason to choose carbon and water, we need to turn to the stars.

The birth of a star – and their planets – starts in places called molecular clouds; they are made of the leftovers of stars in their final stages of evolution (Fig. 1.1a). Molecular clouds have two components: dust and gas. Gas is mostly molecular hydrogen (H_2), and the rest is dominated by organic compounds, that is, molecules made of carbon. Dust is made of silicates; these are molecules of silicon, oxygen, and other elements like iron and magnesium. About half of our planet is made of silicates; rocks are silicates. Dust particles have other minor components like inorganic carbon, which is graphite and diamonds smaller than bacteria, called

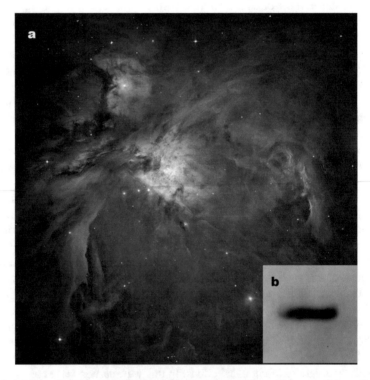

Fig. 1.1 (**a**) Orion Nebula; credit: NASA, ESA, M. Robberto (Space Telescope Science Institute/ESA) and the Hubble Space Telescope Orion Treasury Project Team. (**b**) Edge-on protoplanetary disk in the Orion Nebula; credit: Mark McCaughrean (Max-Planck-Institute for Astronomy), C. Robert O'Dell (Rice University), and NASA/ESA

nanodiamonds. Water is part of the gas and, depending on the temperature, can cover dust particles as ice.

Molecular clouds are no homogeneous; some parts become denser and contract to form a disk structure (Fig. 1.1b). In the middle of the disk, the new star will be formed while the dust in the disk will coagulate to form larger structures; some of them will originate planets. Then, carbon and water in molecular clouds will become part of those planets as the natural process of planetary formation. That is why scientists assume that the chemistry of carbon and the availability of liquid water are our best choices in the current working definition to search for life in other planets. Despite being a single example of life (based on carbon and liquid water), the study of life on Earth is relevant for astrobiology because it (1) gives us the tools to understand the requirements for the origins of life and (2) allows us to understand the most extreme conditions where life can survive. In general, the geological history of our planet is an example of the evolution of a habitable planet, a blue planet where life transformed everything through its metabolism. Astrobiologists consider that extreme conditions in planet Earth are akin to the conditions present in other bodies of the Solar System without the need of using a spaceship. However, it remains to be seen to what extent biological adaptations to those extreme extraterrestrial environments might be similar or different to those observed on Earth, given that most complex adaptations in our planet are the results of a long and contingent evolutionary history.

1.3 The Limits of Life

Earth has hosted living organisms during the past 4.0–3.5 billion years (Nutman et al. 2016; Tashiro et al. 2017). This time has been enough for biological evolution to explore traits that had allowed organisms to colonize every habitat compatible with the stability of biomolecules and where liquid water is available. Since the discovery of microorganisms living in the geothermal hot springs of Yellowstone at temperatures near the boiling point of water and very low acidity values (Brock 2012), we now know that very diverse communities of organisms can thrive in niches with physical or geochemical conditions that exceed those where the more familiar forms of life proliferate. These organisms are called extremophiles as they can grow at harsh conditions of temperature, pH, pressure, dryness, salinity, and even radiation. Extremophiles are defined by the environmental condition in which they grow optimally as is shown in Table 1.1. Extremophiles are primarily prokaryotic, it means single-celled organisms belonging to the Bacteria and Archaea domains, but a few examples are also found in the Eukarya domain. Nevertheless, the most extreme organisms belong to the Archaea domain. Extreme environments are associated with deserts, hydrothermal vents, springs, acid mine drainages and rivers, permafrost sediments, alkaline lakes, and nuclear reactors, among others (Mastascusa et al. 2014). Some extremophiles are adapted simultaneously to

Table 1.1 Classification of extremophiles as defined by their environment

Environmental parameter	Type of extremophile	Defining growth condition	Representative extreme environment
Temperature	Psychrophile	$< -15\ °C$	Ice, snow
	Thermophile	60 to 80 °C	Hot springs
	Hyperthermophile	110 to 121 °C	Oceanic hydrothermal vents, hot springs
pH	Acidophile	pH < 3.0	Mining drainage, volcanic springs
	Alkalophile	pH > 9.0	Alkaline lakes
Pressure	Piezophile	> 10 MPa	Deep ocean, e.g., Mariana Trench
Salinity	Halophile	> 10% (w/v) NaCl	Salt lakes, deserts, salt mines, salted food, evaporites
Ionizing radiation	Radioresistant	1500 to 6000 Gy	Nuclear reactor water core
Desiccation	Xerophile	$a_w < 0.6$	Deserts, rock surfaces, hypersaline environments
Heavy metals	Metallotolerant	High concentrations of Cu, Cd, As, Zn	Contaminated waters, soils
Nutrients	Nutritionally limited environments	Oligotroph	Pelagic and deep ocean
Rock-dwelling	Endolith	Resident inside rocks	Upper subsurface to deep subterranean locations

multiple stresses, and they are recognized as polyextremophiles; common examples are the thermoacidophiles or the haloalkaliphiles.

Extremophiles provide good models for the study of biodiversity on Earth; they are relevant for the recognition of the boundaries of life, the formulation of theories about the origin of life, to sustain the search for life in extraterrestrial scenarios, and to get insights into the potential of other worlds to support life, all of which are important issues in astrobiology.

Current environmental and theoretical studies suggest that the upper limit of life might lay near 150 °C, as this is the temperature value where most of the macromolecules used by living organisms, such as the thermophiles and hyperthermophiles, remain functional. Similarly, thermodynamic considerations suggest that life might be impossible below −40 °C (Price and Sowers 2004). Extremophiles that are active at cold temperatures, in saline environments, or at anoxic conditions are of particular interest for astrobiologists. They can provide clues to assess the habitability potential of the liquid water salty oceans suspected to exist below the icy crust of some of the satellites in the outer Solar System, or on the surface and subsurface of Mars, where sporadic episodes of liquid water were detected below the Martian polar caps (Orosei et al. 2018) or as oxychlorinated brines have been reported (Ojha et al. 2015).

Despite ongoing scientific investigations in our planet for most of the recorded human history, we still find life in unexpected places; and given the number of Earth ecosystems that still need to be explored in detail, we expect the current boundary of life to be pushed even further. While considering the possibility for life to originate and exist on other planetary bodies, it is important to understand the variability of Earth's local conditions when compared to the planetary mean. The ability of earthly microorganisms to colonize extreme environments expands the range of extraterrestrial bodies that may be candidates for extant life, or that may have harbored life in the past, and therefore may retain fossil records. Knowledge of survival mechanisms in extremophiles will also assist the understanding of how extraterrestrial life may survive space travel (e.g., in meteorites). Biological studies will also provide important information about potential biomarkers (Cavicchioli 2002). Biomarkers may be biological remnants such as the presence and type of lipid remnants (Summons and Walter 1990), the chirality of amino acids and sugars, and carbon isotope ratios[1] (Sumner 2001). Most parameters considered are unlikely to be extreme over an entire planet, and local or transient conditions might still support life. Whether or not other planetary bodies such as Mars, Enceladus, or Europa could or did support life, the search for Earth's life true limits will inform our exploration of space and could provide insight into processes that have led to the origin of life on our planet (Merino et al. 2019).

1.4   Extreme Sites on Earth

The search for the suitable conditions for life in the planetary bodies of the Solar System can be achieved either by studying specific environments on Earth that exhibit similar conditions as those reported for the planets and satellites in the Solar System (i.e., planetary field analogues) or by recreating planetary conditions in well-controlled laboratory setups (i.e., laboratory analogues). Frequently these approaches are complemented with observations of specific planetary targets.

There are some terrestrial sites with extreme ranges of temperature, salinity, mineral content, pH, pressure, and water availability that have been classified as analogues to any of the planetary bodies of astrobiological importance, such as the rocky planets Venus or Mars, the Jovian moons Europa or Ganymede, and the Saturn satellites Titan and Enceladus (Marlow et al. 2011; Preston and Dartnell 2014). Terrestrial analogues help to optimize scientific and technological needs such as the tests of machinery and scientific instruments, as well as exploration strategies or preparation of astronauts for robotic or manned space missions. A good analogue site does not replicate all the conditions of another planetary object, but instead, it mimics specific parameters such as mineralogy, elemental abundances, organic content, redox potential, water activity, temperature, radiation, mechanical, and bulk physical characteristics (Marlow et al. 2008; Marlow et al. 2011). Besides,

---

[1] Isotopes are atoms of the same element that have different atomic masses due to the number of neutrons in their nuclei. For example, carbon has two stable isotopes ^{12}C with six neutrons and ^{13}C with seven neutrons.

analogue sites can be used while the expensive and time-consuming space exploration missions are prepared.

Among the most representative analogue sites relevant to astrobiology are Rio Tinto in Spain and the cold and hyperarid Atacama Desert in Chile as analogue sites for the present planet Mars, as well as the black smokers in the deep sea or the Antarctic lake system in Antarctica as analogues sites for the icy satellites, or the Alaskan oil fields for Titan's environment. In Mexico, several analogue sites of astrobiological interest for different topics have been identified. A detailed description of them can be found in Chap. 2 of this volume.

On the one hand, several microbiological communities have been identified at the Rio Tinto, in southern Spain, despite its acidic waters having a very high iron concentration. This makes Rio Tinto a good electrochemical and compositional analogue of early Mars, in particular, to study the possible role of microorganisms on the formation of iron oxide and sulfate deposits on the Red Planet (Amils et al. 2007).

On the other hand, the regions of Yungay in Chile and Arequipa in Peru that forms the Atacama Desert are the driest hot places on Earth, serving as a compositional Martian analogue because of its sulfate and perchlorate mineralogy and low organic content (Catling et al. 2010).

Also, the Antarctic subsurface lake system is an environment closely linked to the subsurface liquid water ocean of the satellite Europa. The largest and most comprehensively studied is Lake Vostok located beneath Russia's Vostok Station under 3488 m of ice, with dimensions of 250 km long by 50 km wide, covering an area of 12,500 km^2, and a volume of 5400 km^3. Lake Vostok waters have been isolated for the last 15 to 25 million years, so it has been hypothesized that in this sealed environment, unusual forms of life could be found. The conditions of the Lake could resemble those of the ice-covered oceans of the Jovian satellites Europa and Ganymede and the Saturnian satellite Enceladus (Priscu et al. 1999).

In the vicinity of deep-sea hydrothermal vents, no sunlight reaches (see Chap. 7), so sulfur compounds are utilized as a source of energy, as well as the conversion of CO_2 and H_2 into CH_4 (McCollom 1999). Similar environments might exist on the satellites Europa and Enceladus. Also, sulfate salts have been detected in Europa's icy surface, possibly providing a sample of the ocean underneath (Brown and Hand 2013), and in the plumes of Enceladus (Bouquet et al. 2015).

All these facts indicate that the efforts to continue advancing not only in the recognition of our planet but also with the exploration of the objects in the Solar System and from other planetary systems as possible niches for life must go on.

1.5   Where Are Astrobiologists Looking for Life?

Planets are the places where all the raw materials for building life may be available along with several sources of energy ideal to promote the chemical reactions that life require to emerge and evolve. Carbon is abundant in the Universe, thus we expect it will be part of planets in general, and then the main requirement is liquid water. Potentially habitable planets are those that have liquid water. This strategy is based on the motto "follow the water" (Hays et al. 2015), and its purpose is to identify the physical conditions or evidence that indicates the past or present presence of water in a planetary object. For example, the geological record on Mars includes minerals and structures that required liquid water, although this compound is not liquid anymore on the surface of the Red Planet (e.g., Carr and Bell 2014; Golombek and McSween 2014).

The only habitable planet we know now is Earth, and we use it to understand how a potentially habitable planet may work and change in time. So, astrobiology is not only "astro" and "biology" but geology too. Unlike the case of life, where there is only one example, the case of potentially habitable planets has many examples. In our Solar System, there are eight planets and tens of natural satellites that we can use to make comparative studies with respect to Earth. This comparative method is used in order to understand if there is something special about Earth and if there are other planetary bodies in the Solar System that are or were potentially habitable. Besides, we also have the exoplanets (i.e., planets around other stars) that may also able to sustain life.

1.6   The Closest Neighbors: Planetary Bodies in the Solar System

Exploration of the Solar System started using remote observations, that is, using instruments to measure the movement of planets and telescopes that observe some details of the surface or the atmospheres of planets. By the 1970s, the exploration included spacecrafts that flew by many planets, like the Pioneer 10 and 11 missions, while others were dedicated just to one planet, like the Mariner (USA) and Venera (Soviet Union) missions. Each of these missions has several probes, even when many of those probes could not fulfill their purpose, the successful ones opened the window for a closer look of our Solar System. Venera 7 was the first to send information from the surface of another planet, starting the possibility of studying planets in situ. Venus and Mars were the first planets to be explored, and their characteristics offered clues for understanding planetary habitability.

1.6.1   Venus, Earth's Hot Twin

Venus seemed very similar to Earth because of its size and mass (Table 1.2), but its atmosphere is 92 times denser and composed mostly by carbon dioxide (CO_2). Clouds that cover Venus are made of sulfuric acid instead of water, as the last one is scarce on Venus and is present only in the gaseous phase at high altitudes in the atmosphere. Temperatures over 400 °C at the surface eliminate any possibility of

Table 1.2 Physical, orbital, and atmospheric characteristics with exploration information of some of the astrobiological important objects in the Solar System

	Earth	Venus	Mars	Europa	Titan
General description	Terrestrial planet			Icy satellite of Jupiter	Satellite of Saturn
Discovery		Visible to the naked eye		Galileo Galilei, Simon Marius (1610)	Christiaan Huygens (1655)
Mean distance from Sun (AU)	1.0	0.73	1.52	5.2	9.5
Mass (kg)	5.97×10^{24}	4.87×10^{24}	6.42×10^{23}	4.79×10^{22}	1.34×10^{23}
Equatorial radius (km)	6378	6052	3397	1561	2575
Mean density (g/cm³)	5.51	5.24	3.93	3.01	1.88
Surface gravity (m/s²)	9.81	8.87	3.72	1.31	1.35
Orbital period (days)	365.26	224.70	686.97	3.55	15.95
Rotational period (days)	0.997	−243.025 retrograde	1.026	3.55	15.95
Axial tilt (°)	23.44	2.64	25.19	0.1	0.0
Visual geometric albedo	0.367	0.689	0.170	0.67	0.22
Average surface temperature (K)	288	737	210	102	93.7
Atmospheric major constituents (%)	N_2 (78.08), O_2 (20.95), H_2O_v (1.0), Ar (0.93), CO_2 (0.04)	CO_2 (96.5), N_2 (3.5), SO_2 (0.015), Ar (0.007), H_2O_v (0.002)	CO_2 (95.97), Ar (1.93), N_2 (1.89), O_2 (0.15), CO (0.06)	O_2	N_2 (98.4), CH_4 (1.4), H_2 (0.2), Hydrocarbons, Nitriles
Surface pressure (bar)	1.01	92	6.4×10^{-3}	1×10^{-11}	1.47
Natural satellites	Moon	None	Phobos and Deimos		

(continued)

Table 1.2 (continued)

	Earth	Venus	Mars	Europa	Titan
Liquid water	Oceans, seas, lakes, rivers, groundwater		Oxychlorinated brines in the surface, subsurface	Subsurface, global ocean	No, but liquid methane seas and lakes on the surface. Probable subsurface ammonia-water global ocean
Spacecraft visitors		Venera program (1961)[a], Mariner 2 (1962)[b], Pioneer Venus (1978)[b], Vega 1 and 2 (1985)[a], Magellan (1990)[b], Venus Express (2006)[c], Akatsuki (2010)[d]	More than 20 successful missions. A complete updated list can be consulted in the NASA Mars exploration website[e]	Pioneer 10 and 11 (1973, 1974)[b]. Voyager program (1979)[b], Galileo mission (1996)[b]	Voyager program (1977)[b], Cassini-Huygens (2004)[b, c]

[a]Soviet Union Space Program. [b]NASA: National Aeronautics and Space Agency (USA). [c]ESA: European Space Agency. [d]JAXA: Japan Aerospace Exploration Agency. [e] https://mars.nasa.gov/mars-exploration/missions/historical-log/

liquid water (Taylor and Hunten 2014; Taylor et al. 2018). Therefore, Venus is not an actual habitable planet. However, composition of the surface and atmosphere of Venus provides evidence that once there was liquid water on its surface.

The evolution of Venus from a planet with liquid water to a dry and hot desert is linked to Sun history. Our star, as all the stars that convert hydrogen into helium in their cores (aka main sequence stars), is slowly increasing its luminosity. When the planets formed, the Sun emitted 70% of the energy that emits today. As the Sun luminosity increased, Venus surface temperature raised too. As liquid water in the Venusian surface evaporated (water is a greenhouse gas), the temperature kept on increasing, this is called "runaway greenhouse," and once it starts, temperature and water vapor rise until there is no liquid water left on the surface. From this hypothesis, we would expect a steamy atmosphere, but that is not what we see in Venus. Gaseous water molecules are broken by ultraviolet light, and hydrogen and oxygen atoms escape to space. Thus, even if Venus has not been habitable for a long time, it is a good example of what happens with a potentially habitable planet that is too close to its star.

1.6.2   Mars, the Martians, and the Emergence of Astrobiology

Our red neighbor has been the most important planetary body for astrobiology. The history of Mars exploration goes hand to hand with the first hypothesis and experiments related to the search of life on other planets. On December 9, 1906, *The New York Times* published an article entitled "There is life on the planet Mars" (Whiting 1906). The article presented the research of American astronomer Percival Lowell who was convinced that Mars was inhabited by a civilization that built canals to irrigate the Martian equator. Canals were a scientific fact, according to Lowell. His conclusions were derived from telescope observations that were captured in drawings (Lowell 1895). Thus, human hands, eyes, and brain were responsible for obtaining, registering, and analyzing the evidence, and humans make mistakes. There was controversy about the existence of the canals, and as the observing instruments were improved, the evidence of canals vanished (Sagan and Fox 1975), but not the idea that Mars had life (Salisbury 1962).

Once photographs were used for astronomy, it was evident that Martian surface seasonally changed becoming darker in spring. This led to the idea that there was vegetation on Mars. A Russian astronomer, Gavriil Adrianovich Tikhov (1875–1960), funded the Sector for Astrobotany of the Kazakhstan Academy of Sciences. His idea was that vegetation could be detected in the reflected light of a planet, so he measured the light reflected by several plants and compare them with Mars observations (Sullivan and Carney 2007). He published two books, *Astrobotany* in 1949 and *Astrobiology* in 1953 – this was the first major work to hold the name "astrobiology" (Cockell 2001). In 1962, one of the most important scientific journals, *Science*, published the paper "Martian Biology" where Salisbury proposed a biogeochemical cycle for Mars (Salisbury 1962). By then it was known that the surface temperature of Mars was below zero degrees Celsius most of the time and its atmosphere was tenuous and composed by CO_2 (see Table 1.2). Salisbury's idea was tested by another research group who concluded: "Thus, whatever environments may be encountered during future explorations of the solar system there is every prospect that those which are not too extreme relative to Earth, for example, those expected on Mars may well support forms of life that we could recognize readily" (Siegel et al. 1963).

The first experiment designed and performed to search life on other planet traveled in two of the four spacecrafts from the *Viking* mission; both landers arrived on Mars surface on 1976, *Viking 1* at Chryse Planitia and *Viking 2* at Utopia Planitia (NASA Viking missions). They carried several experiments to study the surface and atmosphere of Mars, three biology experiments to detect different possible metabolic pathways in presumed Martian organisms and an instrument for the search of organic material on the surface. The three experiments happened to be positive, almost. Part of the response in every experiment was consistent with life, but at the same time, some other part of the experiment was negative. Even more, the search for organic material in the Martian soil was negative. The conclusion was that non-biological chemical reactions were the best explanation for the overall results; mainly a highly oxidized chemical compound was responsible for what was observed in the experiments (Jakosky et al. 2007). In 2008 the Phoenix lander (NASA Phoenix Mission) identified perchlorates in the arctic plains of Mars (Hand 2008). Perchlorates (ClO_4^{2-}) are chemical species made of one chlorine and four oxygen atoms; these are highly oxidized compounds. Quinn and collaborators (Quinn et al. 2013) replicated the biology experiments under conditions like those reported on Mars, including perchlorates, and found that reactions with these compounds could explain the results of all Viking biology experiments (see Chap. 2 for more details on the detection of organics in Mars).

1.6.2.1 Martians on the Rocks

For the scientific community, the consensus is that Viking experiments did not found evidence of life on Mars. But this did not close the case for Martians; in fact, for some authors they may be already on Earth, but we just did not see them before because they traveled in rocks instead of spaceships. In 1996, NASA announced an upcoming paper with strong evidence of life on Mars. David McKay and collaborators (1996) analyzed the Martian meteorite ALH84001 and found structures, minerals, and chemical compounds consistent with biological activity. So, let's start with the rock. We have more than a couple of hundreds of rocks on Earth that came from Mars; they are called Martian meteorites (NASA 2012; Baalke). They were expelled from Mars surface by impacts with asteroids and traveled through space until they fall on Earth. We know they came from Mars because the gases trapped in their minerals have the same isotopic signature as those measured in Mars atmosphere by the Viking landers. What is more, we know how much time these Martian meteorites were in space because their surfaces changed due to the constant interaction with high-energy particles called cosmic rays. The oldest of the Martian meteorites is ALH84001, a volcanic rock formed 4 billion years ago in Mars (Lapen et al. 2010), which was ejected from Mars 14 million years ago (Eugster et al. 1997) and was finally falling at Antarctica 13,000 years ago. The rock has carbonates, compounds formed under the action of liquid water, 3.8 billion years old.

The research group led by McKay presented the following evidence: (1) presence of ovoid and tubular structures with sizes of about 100 nm, (2) carbonate globules especially associated to ovoid and tubular structures, (3) ringed carbon molecules called polycyclic aromatic hydrocarbons (PAH), and (4) a magnetic mineral called magnetite, with a shape similar to the fossil remains of magnetotactic bacteria on Earth. The conclusion was: "Although there are alternative explanations for each of these phenomena taken individually, when they are considered collectively, particularly given their spatial association, we conclude that they are evidence for primitive life on early Mars" (McKay et al. 1996). After the paper publication, evidence accumulated in favor of a non-biological origin of the structures and carbon compounds. The agreement among the scientific community has been since then that there are no Martian fossils in AL84001. Despite not finding life on Mars (or in the rock that came from Mars), the Viking experiment and the ALH84001 structures were important because they opened discussions regarding what we should expect as a proof of extraterrestrial life, something for which we have not an agreement yet and that thus remains as an open research in astrobiology.

1.6.2.2 Discovering Mars

Twenty years after Viking, the next successful missions were NASA's *Mars Global Surveyor* (NASA Mars Global Surveyor Mission) and *Pathfinder* (NASA Pathfinder Mission). Mars Global Surveyor obtained high-resolution images and altimeter measurements to map the entire Martian surface. Pathfinder mission carried a robotic rover named *Sojourner* used to study the surface of Mars in an unprecedented way. The rover was able to move around and was directed from Earth to choose the rocks and soil to be photographed and analyzed with its instruments. Viking orbiters and Mars Global Surveyor showed gullies and debris flows produced by a liquid on the surface, Sojourner registered rounded pebbles and cobbles at the landing site, and other observations suggested conglomerates formed by running water.

All the geologic evidence pointed out a very different Mars compared to the present, cold dusty desert. Liquid water on Mars implies a denser carbon dioxide atmosphere that provided higher surface pressure and an enhanced greenhouse effect. From the observation with telescopes, scientists knew there was frozen water in the Martian poles, and missions have found indirect and direct evidence of frozen water beneath the surface of the Red Planet. Putting together the pieces collected by a century of observations, spacecraft and rover missions, we are certain that liquid water was present on Mars surface since 3800 billion years ago. The Red Planet was potentially habitable just by the time life on Earth was arising; even more, both planets shared similar conditions: active volcanoes, liquid water on the surface, an atmosphere of carbon dioxide and nitrogen, and availability of elements such as sulfur and phosphorus, fundamental for making life on Earth. Then, the question is whether life arose on Mars as it did on Earth. If there was life on Mars, what happened? Has it been hiding somewhere in the subsurface? If it was extinguished when or after Mars lost its atmosphere, did it leave some evidence of its presence? We do not have answers to these questions, yet. The European Space Agency (ESA) and NASA have their eyes and instruments on Mars, like *Odyssey* (NASA) and *Mars Express* (ESA), which have been orbiting the Red Planet for more than 10 years. Other countries are joining the Mars exploration, such as India, which successfully launched *Mangalyaan*, an orbiter inserted in Mars orbit in September 2014 and was still operational by September 2019, although it was originally planned to last only 6 months. Finding life on Mars is still an open goal for new missions, as well as the possibility of a human mission (ESA; NASA's Mars Exploration Program).

1.6.3 Life in the Cold: The Icy Satellites

Icy moons circle around the giant planets, Jupiter and Saturn. In general, they are made of combinations of iron, rock, and water. The densest moons, like Io, Europa, and Ganymede, have iron cores and rocky mantles. Over the rocky mantles, Europa and Ganymede have water ices. Other satellites, like Callisto, Titan, and Enceladus, are made of rock and ice only. Despite, being too far from the Sun, to be warm worlds covered by oceans, their interiors may provide the ingredients for the origin and evolution of life.

1.6.3.1 Europa, Ganymede, Callisto, and Enceladus

The Jupiter system has been studied by several missions, but just *Galileo* was dedicated to exploring the Jovian system for 14 years (NASA Galileo Mission). Europa and her sibling satellites Io, Ganymede, and Calisto were discovered by Galileo Galilei and allegedly by Simon Marius at the start of the seventeenth century. Europa may be the best place in the Solar System to look for currently existing life beyond Earth. The Voyager twin spacecrafts were the first to take detailed images of Europa in 1979; they revealed a bright surface crossed with numerous bands, ridges, cracks, and a surprising lack of large impact craters, suggesting that something had erased them. Voyager's data also indicated that Europa's icy surface had slowly migrated eastward with respect to the satellite's tidal axes because a ductile or liquid layer exists between the icy surface and the deeper interior. To complement this explanation, sophisticated theoretical models of tidal heating reinforced the idea of a global subsurface ocean within Europa (Prockter and Pappalardo 2007). The Galileo mission detected a magnetic field induced by the satellite's motion through Jupiter's field, creating eddy currents in a briny water layer beneath the surface, enriched by sulfates and carbonates salts (Chyba and Phillips 2002), and probably with chlorinated compounds (Hanley et al. 2014). Charged particles irradiation of the surface materials can create oxidants that can be transported to the subsurface ocean and could serve as a fuel for simple forms of life (Phillips and Pappalardo 2014). On the other hand, Ganymede, the largest, more massive, and the only satellite in the Solar System that possesses a magnetic field, also hosts an internal liquid water ocean extended by perhaps 800 km. General characteristics of these satellites and other bodies relevant to astrobiology are summarized in Table 1.2. Callisto, the third-largest satellite in the Solar System, shows the oldest and most heavily cratered surface, meaning no signs of geological activity, but leading to partial differentiation and possibly to the formation of a subsurface ocean 100–150 km deep. Because of its low radiation levels, Callisto has been considered a most suitable place for a human base for future exploration of the Jovian system (Troutman et al. 2003). A variety of organic functional groups, such as hydrocarbons, nitriles, or aldehydes, have been detected in these worlds. Carbon dioxide concentrations as high as 0.2 wt% (weight percent) have been also reported, and the accumulation of other organics provided by cometary impacts has been proposed as another source of biogenic elements for these worlds (Chyba and Phillips 2002).

A return mission to Europa is the only way to gather the critical data required to answer the highest priority astrobiological questions about this ocean world. The Europa Clipper mission will fly by the satellite repeatedly observing it with a pay-load specifically designed to address potential habitability, this understood as the ability for a planetary environment to support life forms analogous to known terrestrial ones. The mission consists of an ice-penetrating radar to search for water on the subsurface, an infrared spectrometer to identify molecular compounds, a stereo camera for mapping and topography, a neutral mass spectrometer to identify atmospheric constituents, a magnetometer along with Langmuir probes to measure the induced magnetic field constraining the salinity and thickness of the ocean, and the spacecraft's radio system to undertake gravity measurements. All these instruments would enable the mission to seek evidence of subsurface water chemistry compatible with habitability, and active geological processes driven by tidal flexing and heating (Phillips and Pappalardo 2014).

Around Saturn, a little moon surprised scientists. Enceladus has a radius of 252 km, and it is expelling water and dust from its interior through plumes that feed Saturn's ring E, the largest and most tenuous of the rings. What is amazing is that such a small body has energy sources large enough to produce the observed jets. The output material indicates an environment that may be suitable for the origins of life (McKay et al. 2008) if life can emerge in sites similar to hydrothermal vents (Deamer and Damer 2017). Future missions may sample the plume in search for signatures of life (Tsou et al. 2012).

1.6.3.2 Titan

Titan was discovered in 1655 by the Dutch astronomer Christiaan Huygens. In 1908 the Spanish astronomer Josep Comas i Solà observed a limb darkening, now interpreted as the first evidence of the atmosphere. By 1925 Sir James Dean demonstrated that some of the constituents of the protosolar nebula would not have escaped from Titan gravitational field despite its small size and weak gravity (Table 1.2, Coustenis 2014). In 1944, the Dutch-American astronomer Gerard P. Kuiper detected, by infrared spectroscopy, the strong absorption bands of methane in the atmosphere (Kuiper 1944). *Pioneer 11* was the first probe that took images of Titan by 1979 and determined that it was a very cold place to support life. In 1980, the *Voyager 1* mission made a flyby close to Titan, 4400 km from the surface, and recovered information that allowed the calculation of the density, mass, and composition of the satellite, as well as the temperature of the atmosphere. Unfortunately, atmospheric haze prevented the direct imaging of the surface. Titan has also been observed by ground-based and Earth-bound observatories like the Canada France Hawaii Telescope (CFHT), the Keck, the Very Large Telescope (VLT), the Hubble Space Telescope (HST), and the Infrared Space Observatory (ISO) to extract complementary information from its non-homogeneous surface and the complex organic chemistry developing on its substantial atmosphere. Cassini-Huygens, one of the most ambitious and recent space missions, reached Titan in 2004 to gather both remote sensing and in situ data that has deepened immensely our knowledge of the satellite. The *Cassini* orbiter took the highest-resolution images of Titan's surface during a close flyby at 1200 km, discovering patches of light and dark terrain. The closest flyby at 880 km, in June 2010, allowed the identification of lakes and seas on the surface of the north polar region (Porco et al. 2005). On January 14, 2005, Titan became the most distant body from Earth to have a space probe, *Huygens*, landed on its surface. During the descent, *Huygens* photographed pale hills with dark rivers running down to a dark plain. After landing, a dark plain covered with small rocks and water-ice pebbles was photographed. The surface is dark because it consists of a mixture of water and hydrocarbon ices (Porco et al. 2005). A rediscovering of Titan had begun.

Under a hazy orange atmosphere, there is a hidden world. Dunes extend for kilometers, round white rocks lie over frozen ground, lakes evaporate, clouds are formed, rain lasts for days, and faults and mountains break the surface. All this may sound very similar to Earth, but we need to look deeper. On Titan, the atmosphere is made of molecular nitrogen and methane (Table 1.2, Coustenis 2014), while on Earth the atmosphere is molecular nitrogen and molecular oxygen, with just 1.7 ppm of methane. At a temperature of 180 °C below zero, the surface is made of frozen water; the same is for rocks; in our planet, rocks and surface are made of silicates, a combination of oxygen, silicon, and other metals like iron and magnesium. In Titan, silicates are mixed with water in the core of the satellite. Lakes and clouds are not made of water but methane, ethane, and other hydrocarbons. Methane (CH_4) and nitrogen molecules (N_2) are broken 1000 km over the surface by ultraviolet photons and high-energy particles; then carbon, nitrogen, and hydrogen recombine forming large hydrocarbon molecules that further aggregate to form a haze that covers the entire planet and scatters orange light, all of which produce the global view of Titan. Haze falls on the surface forming dunes. Some of the mountains have craters, and they may be volcanoes where instead of lava, water-ammonia (H_2O-NH_3) mixtures flow. Thus, this is a planetary body with active geology and a constant exchange between the atmosphere and surface. Not only carbon is present, but complex hydrocarbon and nitrile molecules are, and though there is no liquid water, there is methane. Is life possible on Titan?

The major problem for life on Titan is the temperature. Living organisms need a boundary between them and the environment; this ensures their control over the processes needed for obtaining energy, for growing and reproduction. On Earth, the boundaries are cellular membranes made of lipids; these molecules have the property of having hydrophobic and hydrophilic ends, that is, one part that likes water but the other does not. Hydrophobic ends in lipids are the reason for the formation of a small capsule when a drop of oil falls on water. Membranes must be a flexible boundary; on the one hand, they should separate the cell from the environment, but they should also allow nutrients to enter and products of metabolism to exit. What happens at low temperature is that lipids lose their flexibility, they freeze, so that they are not useful to form membranes. Stevenson and collaborators (2015) proposed a membrane composition that may work for this cold world. Instead of long hydro-carbon molecules like lipids, small nitrogen-bearing hydrocarbon molecules may be more appropriate. They ran numerical models that calculated the chemical and physical properties of molecular structures using the most common hydrocarbons with nitrogen found in Titan. The result was the azotosome, a capsule built of acrylonitrile (CH_2CHCN), a molecule made of just seven atoms, three carbon, four hydrogens, and one nitrogen linked using single, double, and triple bonds. For now, this is hypothetical as we are not planning to go back to Titan soon (for more on Titan exploration, see Chap. 2 of this volume).

The temperature in Titan is low at the surface, but underneath, as the pressure increases, the temperature does too. Then, hypotheses for life on Titan include the interior of the satellite planet. In the outermost layers of the core, life could use silicate compounds (Bains 2004); the temperature is below 120 °C (the maximum temperature for the growth of organisms called hyperthermophiles), although high pressures may be a problem, being 10 thousand times the pressure at Earth's surface. Over the silicate core, there is the mantle, a layer made of water ice where fluid inclusions rich in nutrients may exist. Models of Titan's interior predict an aqueous ammonia ocean between the mantle and the crust. In there, organisms may use sulfur compounds for their metabolism. All these niches together have a volume of 4×10^{10} km^3, that is, the double of the terrestrial biosphere (Norman and Fortes 2011). Although carbon-based informational molecules and liquid ammonia would allow – in principle – for evolution to occur by means of natural selection, if there is life on Titan, we expect something other than "life as we know it."

1.7 Habitable Planets Around Other Stars

The Sun is not the only star with planets; we have detected thousands of planets around other stars; we called them extrasolar planets or "exoplanets." Some may be similar to the planets we see in our Solar System, rocky planets like Earth, Mars, and Venus, or icy worlds like Neptune and Uranus, or hydrogen and helium giants as Saturn and Jupiter, but most of the exoplanets we have found have characteristics not seen in our Solar System, which provides a new opportunity for the research of planetary interiors. Yet the question that matters for astrobiology is: are exoplanets habitable?

Unlike planets in our Solar System, spacecrafts cannot explore exoplanets. The closest one circles Proxima Centauri, a star that is 4.2 light years away. Then our only possible strategy to look for life in exoplanets is using telescope observations, which are limited because we cannot observe any details, but only general properties of those planets like their bulk atmospheric composition. Therefore, the strategy is to look for life that has changed their planet on a global scale, so that the planetary surface must be habitable. This means that liquid water must be on the surface which is limited by pressure and temperature conditions, both provided by the atmosphere. While the weight of the atmosphere provides the pressure, greenhouse gases help to keep the energy from the star to warm the planet. Earth, for example, would be frozen without its atmosphere; the greenhouse gases made this planet habitable although too much of them will make it inhospitable for life. Thinking on this we build the concept of the "habitable zone," defined as the region around a star where a rocky planet with atmosphere can maintain liquid water on the surface. A rocky planet is made of an iron core and silicate mantle and crust, just like Earth, Venus, and Mars. This condition is needed because we are assuming there is a surface; planets like Jupiter or Neptune have no surface. Besides, it has been shown that silicate surfaces might have contributed to the organization of organic molecules for the origin of life (e.g., Negron-Mendoza and Ramos-Bernal 2005). Atmospheres on rocky planets are produced by volcanic outgassing: nitrogen (N_2), carbon dioxide (CO_2), hydrogen (H_2), and water. Small planets, like the rocky planets in the Solar System, do not have enough gravity to keep hydrogen molecules, so that their atmospheres will be composed of CO_2 and N_2. Thus, we are looking for habitable planets that had evolved life under particular conditions, these are not habitable planets for humans.

Numerical models calculate the habitable zone for a star with a given luminosity (amount of stellar energy emitted per unit of time), assuming planets with atmospheres made of water, CO_2, and N_2 (Domagal-Goldman and Segura 2013). Once the habitable zone is obtained, we can tell which exoplanets lie in the habitable zone of their star. Of course, this is just the start; we need to know if the planets are rocky and have atmospheres dense enough to keep the surface above the water freezing point. From the observations available, we can derive the planetary radius and/or its mass, but we don't have instruments to detect their atmospheres. Mass and planetary radius constrain the likely composition of a planet, but there is a range of compositions that can result in the observed mass and radius, particularly for planets more massive than Earth and less massive than Neptune, called sub-Neptunes. Those planets may be small versions of Neptune, mini-Neptunes, or big versions of Earth, that is, very large rocks called super-Earths, or ocean planets. Maybe we think of our planet as an ocean planet because it apparently has lots of water, but less than 1% of its mass is water. Ours is a dry rock; for ocean planets 10, 20, or even 50% of their total mass is water. We do not know if such planets may be habitable. Super-Earths, on the other hand, may be habitable because their atmospheres are probably made of hydrogen (H_2), which is used for organisms on Earth and has been hypothesized to sustain life on such exoplanets (Seager et al. 2012; Seager et al. 2013a; Seager et al. 2013b). A short list of exoplanets has been selected as potentially habitable planets by the Planetary Habitability Laboratory (2019), we do not know yet if those planets have atmospheres, or water, or are of rocky composition, their selection is based on the habitable zone calculated by Kopparapu et al. (2013), and assuming that planets smaller than 1.5 Earth radius or 5 Earth masses may be rocky.

1.7.1   A Habitable Planet Around our Closest Stellar Neighbor?

Proxima Centauri is the Sun's closest stellar neighbor; its only detected planet is named Proxima Centauri b; for short, the star is called Proxima and the planet Proxima b. This star is one tenth of the solar mass, and its luminosity is only one thousandth of the solar luminosity; it is classified as an M dwarf or red dwarf. Proxima b is 30% more massive than Earth (Anglada-Escudé et al. 2016), but its radius cannot be measured (Jenkins et al. 2019), so that its composition cannot be constrained (Bixel and Apai 2017), although this has not diminished the interest on Proxima b. The reason is that this exoplanet lies in the middle of Proxima's habitable zone. Proxima b is 0.05 AU (astronomical unit = Sun-Earth distance) from its star; its "year" is of 11 days since this is the time needed to complete one orbit around Proxima. For comparison, Mercury, the Sun's closest planet, is at 0.4 AU from the Sun, and its year lasts 88 days.

What is interesting about this planetary system are the properties of Proxima. As many red dwarfs, this star is "active," which means that the stellar atmosphere produces high-energy particles and radiation capable of breaking molecules (X-rays and extreme ultraviolet = XUV) much larger than those emitted by a star like the Sun. For a planet in the habitable zone of a red dwarf, XUV radiation can heat the planet's upper atmosphere as a result; the atmosphere will flow away from the planet so that there will be no liquid water on the surface without an atmosphere (Luger and Barnes 2015). And it gets worse. Proxima's habitable zone has moved since the planet was formed. All stars start as protostars, when their cores are not hot enough to initiate the conversion of hydrogen into helium; this stage is known as the pre-main sequence. This period lasts a few million years in stars like the Sun, but for less massive stars, it can be extended over one billion years. During this stage the luminosity of the star changes, being larger at the start of the pre-main sequence; therefore, the limits of the habitable zone move. For Proxima the calculated inner limit of the habitable zone started beyond 0.2 AU (Barnes et al. 2018); by the time we suppose Proxima b was formed, the planet was in the zone where it received too much energy for having liquid water at the surface. If Proxima b had water, it was likely lost by the process defined earlier as the runaway greenhouse (see Sect. 1.6.1).. Under this scenario, habitability does not look promising, but there is some hope according to a recent study by Barnes and coauthors (2018). If Proxima b started with a hydrogen atmosphere equivalent to one thousandth Earth masses, then some of that atmosphere may have survived, preserving the planet's habitability potential (Barnes et al. 2018). Meadows et al. (2018a) have calculated what we expect to see with future telescopes for the scenarios predicted by Barnes et al. (2018). So, we can test our hypothesis soon and tell if they were correct or wrong.

1.8   Signatures of Life

Once we have identified habitable planets, we need strategies to recognize the signatures of life. In our Solar System, we can go to the sites and directly analyze the surface and atmosphere and even drill or crash into a planetary body to study what lies beneath the surface to search for present or extant life, or products of life such as chemical compounds or structures (see Chap. 4 for the use of stromatolites as biosignatures). Methane (CH_4) offers an illustrative case of a signature that can be measured on-site (in situ) or with the use of telescopes and shows all the complexity behind the production and identification of a compound as a signature of life.

On Earth, CH_4 is biologically produced by bacteria–called methanogens–living in anoxic extreme environments, such as hydrothermal vents and animal digestive tracts (including humans). Methanogens are important in the formation of microbial mats and stromatolites, since they are also present in anoxic marine and freshwater sediments (Montoya et al. 2011). At least one billion tons of CH_4 is being formed and consumed by microorganisms in a single year (Zheng et al. 2018). However, CH_4 is also abiotically produced by diverse geochemical processes on our planet, such as volcanic degassing and serpentinization at oceanic hydrothermal vents (Martin et al. 2008). Methanogens are evolutionarily restricted to the Archaea domain so far, although a novel light-dependent pathway for methane production using bacterial iron-only nitrogenase has been recently reported on the proteobacteria *Rhodopseudomonas palustris* (Zheng et al. 2018). Similar to biogenic O_2, methane is released to the environment as a waste product from the reduction of low-molecular-weight organic compounds such as CO_2 with H_2, or acetate, or other methylated compounds. This process – known as methanogenesis – is thought to be one of the earliest metabolisms on Earth, very close to the origin of life indeed (Martin et al. 2008; Weiss et al. 2016).

Carbon isotopic composition is essential to understand the origin of methane, and it could also be implemented in planetary landers designed to look for signs of life. Biogenic methane usually, but not always, contains a higher percentage of the lighter carbon isotope (^{12}C), while abiotic methane contains relatively more of the heavier isotope (^{13}C). Because methanogens require oxygen-free (anoxic) conditions and a low redox potential for growth and activity, methanogenesis will occur on modern Earth when other electron acceptors such as oxygen, nitrate, and sulfate are depleted biotically or abiotically from the environment. Under pervasive extreme conditions (such as high salinity, temperature, or pH), methanogens can further face bioenergetic constraints to use less or non-competitive substrates to produce methane (with different carbon isotopic composition). For instance, in hypersaline environments, such as the ones located at Guerrero Negro's salt flats in Baja California or at the sulfate-rich sediments of Tirez Lagoon in Spain, the diversity of substrates and phylogenetic types (phylotypes) producing methane is considerably reduced and bioenergetically constrained within methanogenic archaea (Lozada-Chávez et al. 2009; Montoya et al. 2011, Chap. 8 in this volume). Also, carbon isotopic boundaries of biogenic methane are found to change with salinity differences in the microbial mats, as a probable consequence of methanogens tolerating different salinities by using different substrates (Potter et al. 2009). Thus, identifying life diversity and characterizing their geochemical profiles on extreme environments will help astrobiologists to understand the environmental contexts for false positive and negative mechanisms of biogenic O_2/CH_4 production on possible habitable worlds.

With such goal in mind, microbiologist Lilia Montoya and evolutionary biologist Irma Lozada-Chávez investigated the bioenergetic constraints of methanogenesis under hypersaline conditions at the athalassohaline[2] and sulfate-rich sediment of Tirez lake in Spain, an analogous metabolic system for Mars and Europa (Montoya et al. 2011). For such a task, sediment samples were taken at different depths from winter and summer seasons, their physicochemical parameters were measured, and DNA was extracted from sediments (a very tough task to perform!). They sequenced two functional gene markers that encode highly conserved enzymes used to produce methane (mcrA, in Archaea) and sulfate (aprA, in Archea and Bacteria) to assess the phylogenetic diversity, abundance, and metabolic capacity at Tirez lake. In contrast to ribosomal genes, these functional genes work as metabolic fingerprints (from a huge pool of DNA sequences), by providing an unambiguous link between the genetic identity of an uncultured microorganism and its metabolic activity (Lozada-Chávez et al. 2009; Alvarado et al. 2014; Watanabe et al. 2016). They found few but constant phylogenetic types performing methanogenesis and sulfate reduction/oxidation at Tirez lake. Strikingly, high salt and sulfate concentrations impose a competition between methanogens and sulfate-reducers that use H_2 and acetate as common substrates. By using non-competitive substrates (e.g., methylamine and methanol), other methanogenic phylotypes are not only able to coexist with sulfate-reducers in such extreme conditions, but they also are potentially producing from twice to six times more energy than their counterparts. According to other studies (e.g., Potter et al. 2009), such extra energy is likely used by these methanogens to maintain osmotic balance, which allows them in turn to tolerate higher salinities of sodium chloride (NaCl). These findings have motivating astrobiological implications (now more than ever) since rich salts of sulfate and NaCl have been recently confirmed on the surface of Europa (Trumbo et al. 2019).

1.8.1 Fingerprints of Life in Exoplanets

We can tell if a planet is habitable using its spectrum. The amount of energy emitted or reflected (Fig. 1.2) by a planet in different wavelengths (colors for visible light) tell us how much light reflects the surface and the composition of the atmosphere, if there is one. Another possibility is to use the transmitted spectra, which is the light

[2] An athalassohaline lake is a saline lake not of marine origin, but from evaporation of freshwater in a system dominated by calcium, magnesium, and sulfate (as opposed to sodium and chloride in the ocean).

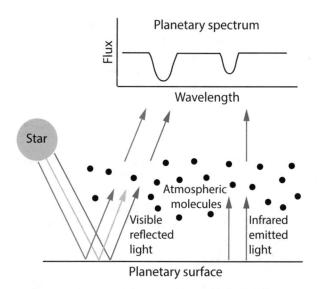

Fig. 1.2 Reflected and emitted planetary spectrum. The spectrum forms when the light emitted or reflected by the planetary surface is absorbed by the atoms or molecules in the atmosphere

of the star that crosses the planetary atmosphere when the planet is in front of the star from our point of view. If there is an atmosphere, specific wavelengths will be absorbed by the atmosphere depending on the atmospheric composition. Each atom and molecule emits or absorbs energy in specific wavelengths that can be used to identify it in a spectrum. Then, we can tell a planet is potentially habitable if we identify carbon dioxide and water vapor in its spectrum. But a habitable planet is not necessarily inhabited, for identifying life on a planet we need a biosignature or a phone call, more specifically a radio call.

A biosignature is a pattern in a planetary spectrum that results from a biological agent (Schwieterman et al. 2018). In particular, we are interested on gaseous biosignatures which result from metabolic activity. Our approach is to focus on the "inputs" and "outputs" regardless the specific biochemical machinery that uses those inputs and produces the outputs. This is based on the fact that life on Earth harnesses energy from chemical reactions where we know the reactants and the products (e.g., Seager et al. 2012). We can find our best example of a biosignature in our planet. When Earth was formed, volcanic outgassing generated an atmosphere of CO_2, N_2, and H_2O. More than 3.8 billion years ago, life emerged on Earth, and around 2.2 billion years ago after a global glaciation, life made the most dramatic global change in the history of this planet. Unicellular microorganisms called cyanobacteria recollect light with pigments and use it to convert CO_2 and H_2O into glucose; the leftover of this process known as photosynthesis is oxygen (O_2). These photosynthetic organisms contaminated Earth for the first time producing a global climatic change. No doubt that cyanobacteria initiated big changes on our planet; from an oxygen-free atmosphere, we ended today with an atmosphere containing

21% of oxygen. These organisms are still producing oxygen as well as plants do. If they were extraterrestrials looking at our planet's spectrum, they could clearly see the water and oxygen absorption bands (Fig. 1.3).

Therefore, for most astrobiologists (e.g., Meadows et al. 2018b), oxygen is considered the strongest biosignature so far, due to the following:

(a) Reliability: O_2 has been produced by life on Earth since at least 3.7 Ga, according to the stromatolites oldest fossils (Nutman et al. 2016), and has no significant geological sources.
(b) Survivability: O_2 has risen over time to become the second most abundant gas in our atmosphere (~ 21% by volume), with major biological consequences at planetary scale, including the evolution of complex multicellular life (Lozada-Chávez et al. 2011) and a biomass dominance of land plants on Earth (~80% of 550 gigatons of Earth's biomass of carbon), which strikingly contrast the biomass of bacteria (~15%) and archaea (~1.3%) being produced mainly at deep-anoxic subsurface environments (Bar-On et al. 2018).
(c) Detectability: O_2 is accessible to distant observation by several methods, including transit spectroscopy. However, oxygen is not the only biosignature worth astrobiological investigation. With the identification of methane (CH_4) in the atmospheres of Mars and Titan, as well as in the Enceladus plumes, CH_4 has gained astrobiological interest as another potential biosignature gas (Young et al. 1997; Taubner et al. 2018).

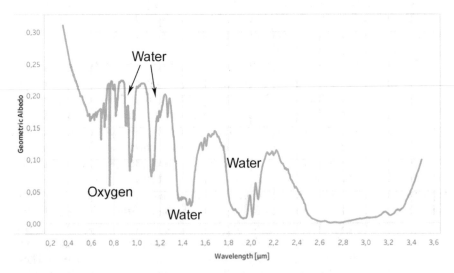

Fig. 1.3 Earth spectrum. Absorption of oxygen (O_2) is evident at 0.76 microns (μm), while water band absorption dominates most of the visible (0.35 μm to 0.7 μm) and near-infrared (0.7 μm to 5 μm). Data from Robinson et al. (2011), image from the VPL Spectral Explorer (http://depts.washington.edu/naivpl/content/vpl-spectral-explorer)

The question is if a potential biosignature can have false positives, that means it could be produced by non-biological processes that we have not identified. Oxygen, for example, is present in Mars and Venus atmospheres as a result of the breakup of their most abundant atmospheric gas, CO_2, by ultraviolet (UV) light. The difference between that oxygen and ours is that we can predict the abundance of that O_2 from a given amount of CO_2 and UV light flux. Then, for our planet the only explanation for the 21% O_2 filling its atmosphere is life; this implies that unicellular life can globally change a planet such as to be detectable with a telescope. We have created a strategy to detect unicellular life with a telescope.

The strategy sounds straightforward, but it is not. We have identified just a few compounds that may work as biosignatures (e.g., Domagal-Goldman et al. 2011; Seager et al. 2012; Schwieterman et al. 2018), and there is another caveat, planetary spectrum won't be anything like the one in Fig. 1.3. Instruments to characterize the atmosphere of potentially habitable planets are not yet available. Once they become available, the measured spectra won't be as detailed as the ones we calculated with models. On top of the fuzzy observations, here we have the same problem as in the Martian meteorite or the Viking experiments, how will we distinguish a biologically produced signature from one that was created from other processes? This is a key problem for astrobiologists. Current research in biosignatures to answer this question is focused on (1) understanding false positives for life, that is, planetary processes that may produce detectable amounts of compounds that can be confused with those produced by living things, and (2) generating observational strategies to distinguish between both scenarios (Meadows 2017; Catling et al. 2018; Kiang et al. 2018; Fujii et al. 2018; Walker et al. 2018).

1.9 What About Intelligent Life?

We may expect something more spectacular than gases produced by bacteria or hollow prints left by extinct life. Although that would be a breakthrough, we may have the fantasy of communicating with alien civilizations. Astrobiologists have that dream too. In the late 1950s, Frank Drake had recently earned a Ph.D. and was working at the National Radio Astronomy Observatory (NRAO) located at West Virginia, United States. Radio telescopes are antennas that detect and emit light with wavelengths from centimeters to meters. Radio waves travel fast, and they are not absorbed by the dust between the stars. Also, Earth's atmosphere is transparent to radio waves making easy to detect them from the ground. Then, Drake had the idea of searching for signals of extraterrestrial (ET) civilizations with the telescopes at Green Bank that were under construction. He was granted permission by the interim director, as long as there was no press involved. Just before starting the project OZMA, the first campaign to search for ET signals, two high energy physicists at Cornell University published a paper proposing the use of radiofrequencies for interstellar communications with other civilizations (Cocconi and Morrison 1959). They started the manuscript by stating that there were no theories to calculate if life may emerge on other planets, but they made a good point at the end by saying: "The probability of success is difficult to estimate; but if we never search, the chance of success is zero."

Certainly, until now, such theory does not exist, but we can organize our ignorance. In 1961, Frank Drake proposed an equation that has been used to calculate the number of technological civilizations in our galaxy. The "Drake Equation," as it is known, does not actually provide an exact number but helps us to identify all the factors that may be involved in the development of other civilizations (SETI Institute 2019). The Equation has been modified to include different factors, for example, Sara Seager, professor at the Massachusetts Institute of Technology, changed the Equation to calculate how many planets around low-mass stars may have detectable biosignatures (Seager 2018).

At first, the efforts for searching extraterrestrial civilizations were called SETI: Search for Extraterrestrial Intelligence. This name may be misleading. We do not have a definition for intelligence, neither a strategy to detect it remotely. What we can do is to search for extraterrestrial technologies that can either be artifacts, particles, or electromagnetic waves (Tarter 2001, 2007). Since project OZMA, several searches have been conducted by professional and amateur astronomers; published reports may be accessed at https://technosearch.seti.org/. Until now, we have not found evidence of alien technologies, but the search has just started.

1.10 Our Habitable Planet

For the first time in the history of humanity, we can perform experiments that may lead us to find life in other planets or moons. Larger telescopes on the ground and in space will focus on potentially habitable planets, and spacecrafts will continue to explore the bodies of our Solar System. For now, there is only one habitable planet we know of: Earth. As the Sun increases its luminosity as part of its natural evolution, Earth will lose its oceans and become not habitable for any kind of life; that future is millions of years ahead. For now, and the close future, no matter what we do, Earth will still be habitable in the broad sense. Global warming will extinguish many species, but others will survive and thrive. We have learned that there is life that will be happy in hot waters or rivers contaminated by mining products. A revision of Earth's history shows us that life survived when our planet was totally frozen and after the impact of large rocks from the space. Human activity may be only one more of those catastrophes where thousands of species will disappear and new will emerge. Life, as the phenomenon that has dominated Earth for at least 3.5 billion years, won't vanish because of us.

But we should be worried anyway, although the fate of life is not on our hands, the destiny of humanity is. Global warming and pollution do influence human life, and we do not have any other place to go. This is our only home, and we can either use our knowledge to keep it habitable for a few million years more or accelerate our extinction. The Universe does not care.

References

Alvarado A et al (2014) Microbial trophic interactions and mcrA gene expression in monitoring of anaerobic dige–sters. Front Microbiol 5. https://doi.org/10.3389/fmicb.2014.00597

Amils R et al (2007) Extreme environments as Mars terrestrial analogs: The Rio Tinto case. Planetary and Space Science 55(3):370–381

Anglada-Escudé G et al (2016) A terrestrial planet candidate in a temperate orbit around Proxima Centauri. Nature 536:437

Baalke R Mars Meteorite Home Page (JPL). https://www2.jpl.nasa.gov/snc/. Accessed 20 Oct 2019

Bains W (2004) Many chemistries could be used to build living systems. Astrobiology 4:137–167

Barnes R et al. (2018) The habitability of Proxima Centauri b I: evolutionary scenarios. ArXiv: 160806919 [astro-ph]

Bar-On YM et al (2018) The biomass distribution on earth. Proc Natl Acad Sci 115:6506

Benner SA (2010) Defining life. Astrobiology 10(10):1021–1030

Bixel A, Apai D (2017) Probabilistic constraints on the mass and composition of Proxima b. Astrophys J 836:L31

Bouquet A et al (2015) Possible evidence for a methane source in Enceladus' ocean. Geophys Res Lett 42:1334–1339

Brock TD (2012) Thermophilic microorganisms and life at high temperatures. Springer Science & Business Media

Brown ME, Hand KP (2013) Salts and radiation products on the surface of Europa. Astron J 145:110

Carr MH, Bell JF (2014) Chapter 17 - Mars: surface and interior. In: Spohn T et al (eds) Encyclopedia of the solar system, 3rd edn. Elsevier, Boston, pp 359–377

Catling DC et al (2010) Atmospheric origins of perchlorate on Mars and in the Atacama. J Geophys Res Planets 115

Catling DC et al (2018) Exoplanet biosignatures: a framework for their assessment. Astrobiology 18:709–738

Cavicchioli R (2002) Extremophiles and the search for extraterrestrial life. Astrobiology 2:281–292

Chyba CF, Phillips CB (2002) Europa as an abode of life. Orig Life Evol Biosph 32:47–67

Cleland CE (2012) Life without definitions. Synthese 185(1):125–144

Cocconi G, Morrison P (1959) Searching for interstellar communications. Nature 184:844

Cockell CS (2001) "Astrobiology" and the ethics of new science. Interdiscip Sci Rev 26:90–96

Coustenis A (2014) Chapter 38 - titan. In: Spohn T et al (eds) Encyclopedia of the solar system, 3rd edn. Elsevier, Boston, pp 831–849

Deamer D, Damer B (2017) Can life begin on Enceladus? A perspective from hydrothermal chemistry. Astrobiology 17:834–839

Domagal-Goldman SD, Segura A (2013) Exoplanet climates. In: Mackwell SJ et al. (eds) comparative climatology of terrestrial planets. Pp 121–135

Domagal-Goldman SD et al (2011) Using biogenic sulfur gases as remotely detectable biosignatures on anoxic planets. Astrobiology 11:419–441

ESA The ExoMars programme 2016-2020. In: Robot. Explor. Mars. https://exploration.esa.int/web/mars/-/46048-programme-overview. Accessed 20 Oct 2019

Eugster O et al (1997) Ejection times of Martian meteorites. Geochim Cosmochim Acta 61:2749–2757

Fujii Y et al (2018) Exoplanet biosignatures: observational prospects. Astrobiology 18:739–778

Golombek MP, McSween HY (2014) Chapter 19 - Mars: landing site geology, mineralogy, and geochemistry. In: Spohn T et al (eds) Encyclopedia of the solar system, 3rd edn. Elsevier, Boston, pp 397–420

Hand E (2008) Perchlorate found on Mars. Nature. https://doi.org/10.1038/news.2008.1016

Hanley J et al (2014) Reflectance spectra of hydrated chlorine salts: the effect of temperature with implications for Europa. J Geophys Res Planets 119:2370–2377

Hays L et al. (2015) NASA astrobiology strategy

Hermida M (2016) Life on earth is an individual. Theory Biosci 135(1–2):37–44

Hubbart SG (2015) What is astrobiology? In: NASA. http://www.nasa.gov/feature/what-is-astro-biology. Accessed 5 Oct 2019

INEGI (2015) Encuesta sobre la Percepción Pública de la Ciencia y la Tecnología (ENPECYT) 2015. https://www.inegi.org.mx/programas/enpecyt/2015/. Accessed 5 Oct 2019

Jakosky BM et al (2007) Mars. In: Sullivan WI, Baross JA (eds) Planets and life: the emerging science of astrobiology. Cambridge University Press, Cambridge, UK, pp 357–387

Jenkins JS et al (2019) Proxima Centauri b is not a transiting exoplanet. Mon Not R Astron Soc 487:268–274

Kiang NY et al (2018) Exoplanet biosignatures: at the Dawn of a new era of planetary observations. Astrobiology 18:619–629

Kopparapu RK et al (2013) Habitable zones around Main-sequence stars: new estimates. Astrophys J 765:131

Kuiper GP (1944) Titan: a Satellite with an Atmosphere. Astrophys J 100:378

Lapen TJ et al (2010) A younger age for ALH84001 and its geochemical link to Shergottite sources in Mars. Science 328:347–351

Lowell P (1895) Mars. The canals I. Pop Astron 2:255

Lozada-Chávez I et al (2009) Metanogenic diversity through mcrA gene in hypersaline conditions. In: Origins of life and evolution of biospheres, vol 39, pp 382–383

Lozada-Chávez I et al (2011) "Hypothesis for the modern RNA world": a pervasive non-coding RNA-based genetic regulation is a prerequisite for the emergence of multicellular complexity. Orig Life Evol Biospheres 41:587–607

Luger R, Barnes R (2015) Extreme water loss and abiotic O2 buildup on planets throughout the habitable zones of M dwarfs. Astrobiology 15:119–143

Mariscal C, Doolittle WF (2018) Life and life only: a radical alternative to life definitionism. Synthese:1–15

Mariscal C, Fleming L (2018) Why we should care about universal biology. Biol Theory 13(2):121–130

Marlow JJ et al (2008) Mars on earth: soil analogues for future Mars missions. Astron Geophys 49:2.20–2.23

Marlow JJ et al (2011) Organic host analogues and the search for life on Mars. Int J Astrobiol 10:31–44

Martin W et al (2008) Hydrothermal vents and the origin of life. Nat Rev Microbiol 6:805–814

Mastascusa V et al (2014) Extremophiles survival to simulated space conditions: an astrobiology model study. Orig Life Evol Biospheres 44:231–237

McCollom TM (1999) Methanogenesis as a potential source of chemical energy for primary biomass production by autotrophic organisms in hydrothermal systems on Europa. J Geophys Res Planets 104:30729–30742

McKay DS et al (1996) Search for past life on Mars: possible relic biogenic activity in Martian meteorite ALH84001. Science 273:924–930

McKay CP et al (2008) The possible origin and persistence of life on Enceladus and detection of biomarkers in the plume. Astrobiology 8:909–919

Meadows VS (2017) Reflections on O2 as a biosignature in exoplanetary atmospheres. Astrobiology 17:1022–1052

Meadows VS et al (2018a) The habitability of Proxima Centauri b: environmental states and observational discriminants. Astrobiology 18:133–189

Meadows VS et al (2018b) Exoplanet biosignatures: understanding oxygen as a biosignature in the context of its environment. Astrobiology 18:630–662

Merino N et al (2019) Living at the extremes: extremophiles and the limits of life in a planetary context. Front Microbiol 10:780

Mix LJ (2015) Defending definitions of life. Astrobiology 15(1):15–19

Montoya L et al (2011) The sulfate-rich and extreme saline sediment of the ephemeral Tirez lagoon: a biotope for Acetoclastic sulfate-reducing bacteria and Hydrogenotrophic methanogenic archaea. Int J Microbiol 2011:1–22

NASA (2012) Martian Meteorite Compendium. In: Astromaterials Acquis. Curation Off. https://curator.jsc.nasa.gov/antmet/mmc/introduction.cfm. Accessed 20 Oct 2019

NASA (2018) NASA Astrobiology Institute. https://nai.nasa.gov/about/. Accessed 5 Oct 2019

NASA Galileo Mission. In: NASA Sol Syst Explor https://solarsystem.nasa.gov/missions/galileo/overview. Accessed 20 Oct 2019a

NASA Mars global surveyor Mission. In: NASA's Mars Explor. Program. https://mars.nasa.gov/mars-exploration/missions/mars-global-surveyor. Accessed 20 Oct 2019b

NASA Pathfinder Mission. In: NASA's Mars Explor Program https://mars.nasa.gov/mars-exploration/missions/pathfinder. Accessed 20 Oct 2019c

NASA Phoenix Mission. In: NASA's Mars Explor. Program. https://mars.nasa.gov/mars-exploration/missions/phoenix. Accessed 19 Oct 2019d

NASA Viking missions. In: NASA's Mars Explor. Program. https://mars.nasa.gov/mars-exploration/missions/viking-1-2. Accessed 19 Oct 2019e

NASA's Mars Exploration Program. In: NASA's Mars Explor. Program. https://mars.nasa.gov/. Accessed 20 Oct 2019

Negron-Mendoza A, Ramos-Bernal S (2005) The role of clays in the origin of life. In: Seckbach J (ed) Origins: genesis, evolution and diversity of life. Springer, Netherlands, pp 181–194

Norman LH, Fortes AD (2011) Is there life on … titan? Astron Geophys 52:1.39–1.42

Nutman AP et al (2016) Rapid emergence of life shown by discovery of 3,700-million-year-old microbial structures. Nature 537:535–538

Ojha L et al (2015) Spectral evidence for hydrated salts in recurring slope lineae on Mars. Nat Geosci 8:829–832

Orosei R et al (2018) Radar evidence of subglacial liquid water on Mars. Science 361:490–493

Phillips CB, Pappalardo RT (2014) Europa clipper Mission concept: exploring Jupiter's ocean moon. EOS Trans Am Geophys Union 95:165–167

Planetary Habitability Laboratory (2019) The Habitable Exoplanets Catalog. http://phl.upr.edu/projects/habitable-exoplanets-catalog. Accessed 1 Nov 2019

Porco CC et al (2005) Imaging of Titan from the Cassini spacecraft. Nature 434(7030):159–168

Potter EG et al (2009) Isotopic composition of methane and inferred methanogenic substrates along a salinity gradient in a hypersaline microbial mat system. Astrobiology 9:383–390

Preston LJ, Dartnell LR (2014) Planetary habitability: lessons learned from terrestrial analogues. Int J Astrobiol 13:81–98

Price PB, Sowers T (2004) Temperature dependence of metabolic rates for microbial growth, maintenance, and survival. Proc Natl Acad Sci 101:4631–4636

Priscu JC et al (1999) Geomicrobiology of subglacial ice above Lake Vostok, Antarctica. Science 286:2141–2144

Prockter LM, Pappalardo RT (2007) Chapter 23 - Europa. In: McFadden L-A et al (eds) Encyclopedia of the solar system, 2nd edn. Academic Press, San Diego, pp 431–448

Quinn RC et al (2013) Perchlorate radiolysis on Mars and the origin of Martian soil reactivity. Astrobiology 13:515–520

Robinson TD et al (2011) Earth as an extrasolar planet: earth model validation using EPOXI earth observations. Astrobiology 11:393–408

Sagan C, Fox P (1975) The canals of Mars: an assessment after mariner 9. Icarus 25:602–612

Salisbury FB (1962) Martian biology: accumulating evidence favors the theory of life on Mars, but we can expect surprises. Science 136:17–26

Schwieterman EW et al (2018) Exoplanet biosignatures: a review of remotely detectable signs of life. Astrobiology 18:663–708

Seager S (2018) The search for habitable planets with biosignature gases framed by a 'Biosignature Drake Equation.'. Int J Astrobiol 17:294–302

Seager S et al (2012) An astrophysical view of earth-based metabolic biosignature gases. Astrobiology 12:61–82

Seager S et al (2013a) A biomass-based model to estimate the plausibility of exoplanet biosignature gases. Astrophys J 775:104

Seager S et al (2013b) Biosignature gases in H_2-dominated atmospheres on rocky exoplanets. Astrophys J 777:95

SETI Institute (2019) Drake Equation. https://www.seti.org/drake-equation-index. Accessed 19 Dec 2019

Siegel SM et al (1963) Martian biology: the Experimentalist's approach. Nature 197:329–331

Stevenson J et al (2015) Membrane alternatives in worlds without oxygen: creation of an azotosome. Sci Adv 1:e1400067

Sullivan W, Carney D (2007) History of astrobiological ideas. In: Sullivan WI, Baross JA (eds) Planets and life: the emerging science of astrobiology. Cambridge University Press, Cambridge, UK, pp 9–45

Summons RE, Walter MR (1990) Molecular fossils and microfossils of prokaryotes and protists from Proterozoic sediments. Am J Sci 290(A):212–244

Sumner DY (2001) Microbial influences on local carbon isotopic ratios and their preservation in carbonate. Astrobiology 1:57–70

Tarter J (2001) The search for extraterrestrial intelligence (SETI). Annu Rev Astron Astrophys 39:511–548

Tarter JC (2007) Searching for extraterrestrial intelligence. In: Sullivan WI, Baross J (eds) Planets and life: the emerging science astrobiology. Cambridge University Press, Cambridge, UK, pp 513–536

Tashiro T et al (2017) Early trace of life from 3.95 Ga sedimentary rocks in Labrador, Canada. Nature 549(7673):516–518

Taubner R-S et al (2018) Biological methane production under putative Enceladus-like conditions. Nat Commun 9:1–11

Taylor FW, Hunten DM (2014) Chapter 14 - Venus: atmosphere. In: Spohn T et al (eds) Encyclopedia of the solar system, 3rd edn. Elsevier, Boston, pp 305–322

Taylor FW et al (2018) Venus: the atmosphere, climate, surface, interior and near-space environment of an earth-like planet. Space Sci Rev 214:35

Trifonov EN (2011) Vocabulary of definitions of life suggests a definition. J Biomol Struct Dynamics 29(2):259–266

Troutman PA et al (2003) Revolutionary concepts for human outer planet exploration (HOPE). AIP Conf Proc 654:821–828

Trumbo SK et al (2019) Sodium chloride on the surface of Europa. Sci Adv 5(6):eaaw7123

Tsokolov SA (2009) Why is the definition of life so elusive? Epistemological considerations. Astrobiology 9:401–412

Tsou P et al (2012) LIFE: LIFE investigation for EnceladusA sample return Mission concept in search for evidence of LIFE. Astrobiology 12:730–742

Walker SI et al (2018) Exoplanet Biosignatures: Future Directions. Astrobiology 18:779–824

Watanabe T et al (2016) Identity of major sulfur-cycle prokaryotes in freshwater lake ecosystems revealed by a comprehensive phylogenetic study of the dissimilatory adenylylsulfate reductase. Sci Rep 6:1–9

Weiss MC et al (2016) The physiology and habitat of the last universal common ancestor. Nat Microbiol 1:1–8

Whiting L (1906) There is life on the planet Mars N Y Times 1

YouGov (2015) You are not alone: most people believe that aliens exist. https://yougov.co.uk/topics/lifestyle/articles-reports/2015/09/24/you-are-not-alone-most-people-believe-aliens-exist. Accessed 5 Oct 2019

Young LA et al (1997) Detection of gaseous methane on Pluto. Icarus 127:258–262

Zheng Y et al (2018) A pathway for biological methane production using bacterial iron-only nitrogenase. Nat Microbiol 3:281–286

Chapter 2
Astrobiology and Planetary Sciences in Mexico

Karina Cervantes de la Cruz, Guadalupe Cordero-Tercero,
Yilen Gómez Maqueo Chew, Irma Lozada-Chávez, Lilia Montoya,
Sandra Ignacia Ramírez Jiménez, and Antígona Segura

Abstract A small community of scientists in Mexico has been contributing to the study of planetary bodies in our Solar System and around other stars, including their potential for habitability. Here, we present particular aspects of this research told as a journey: from the first attempts to reproduce cells and the laboratories where the first Mexican astrobiologists were educated to the sites in Mexico where scientists are studying the extremes of life and likely environments of other planets. We jump to space rocks that narrate the history of the Solar System. Then, we move to Mars and the debate of organics and the Viking experiment to continue with the hidden water oceans of the icy satellites and Titan, an exotic orange satellite with methane lakes, hydrocarbon dunes, and water ice rocks. Our journey continues toward other

---

All authors contributed equally to the text, listed in alphabetical order.

K. Cervantes de la Cruz
Departamento de Física, Facultad de Ciencias, Universidad Nacional Autónoma de México,
Mexico City, Mexico

G. Cordero-Tercero
Instituto de Geofísica, Universidad Nacional Autónoma de México, Mexico City, Mexico

Y. Gómez Maqueo Chew
Instituto de Astronomía, Universidad Nacional Autónoma de México, Mexico City, Mexico

I. Lozada-Chávez
Interdisciplinary Center for Bioinformatics, University of Leipzig, Leipzig, Germany

L. Montoya
Universidad Autónoma del Estado de Morelos, Cuernavaca, Morelos, Mexico

S. I. Ramírez Jiménez (✉)
Centro de Investigaciones Químicas, Universidad Autónoma del Estado de Morelos,
Cuernavaca, Morelos, Mexico
e-mail: ramirez_sandra@uaem.mx

A. Segura
Instituto de Ciencias Nucleares, Universidad Nacional Autónoma de México, UNAM,
Mexico City, Mexico

© Springer Nature Switzerland AG 2020
V. Souza et al. (eds.), *Astrobiology and Cuatro Ciénegas Basin as an Analog of Early Earth*, Cuatro Ciénegas Basin: An Endangered Hyperdiverse Oasis,
https://doi.org/10.1007/978-3-030-46087-7_2

stars where we search for planets beyond our Solar System, known as exoplanets, that have shown a surprising diversity more familiar to science fiction with hot Jupiters, lava worlds, mini-Neptunes, super-Earths, and potentially habitable worlds.

2.1 Astrobiology and Planetary Sciences in Mexico

The exploration of other worlds is a young endeavor, telescopes and meteorites were our only chance to study other planets until the first missions were launched and successfully orbited and landed other bodies of our Solar System after the 1960s. Astrobiology is even younger, consolidating its purpose and methods during the last decade of the past century (see Chap. 1 in this volume). But for both, planetary sciences and astrobiology, their roots can be traced from Earth sciences, astronomy, and the study of the origin of life. After all, the planet and life we have on hand to study are our own. Mexico has a long tradition in those sciences. Here we will review part of that history and the most recent projects lead by Mexican scientists that are contributing to the understanding of the Solar System, planets around other stars, and life as a universal phenomenon.

2.1.1 Plasmogenesis: A Hypothesis for the Origins of Life

Alfonso Luis Herrera was a dangerous man. He was convinced that life was the result of a mixture of chemical compounds that could be reproduced in a laboratory (Herrera 1919). He was the author of the first biology textbook in Mexico and one of the main promoters of Darwinism in this country at the start of the twentieth century (Ledesma-Mateos and Barahona 2003). In 1902, Herrera established the first general biology course in Mexico at the Escuela Normal, a school dedicated to teachers' education (Ledesma-Mateos and Barahona 2003). These activities were noticed by powerful politicians and religious people leading to the cancellation of the course in 1906 because it was considered "dangerous to the young people and to beliefs" (Ledesma-Mateos and Barahona 2003).

Plasmogenesis was born from Herrera's idea that the functions and structures of living beings could be explained from physicochemical laws (Negrón-Mendoza 1995). He proposed this new science in 1904 and dedicated 30 years to experiments focused on "producing artificial cells intended to mimic as perfectly possible the structures of nature." (Herrera 1919). In 1942, he reported the formation of "sulphobes," cell-like microstructures formed by evaporation of mixtures of inorganic compounds (Negrón-Mendoza 1995). His approach to the origins of life prevails until today and precedes Oparin's book, The Origin of Life (Oparin 1924) and Miller's experiments (Miller 1953). For astrobiology, this view of life means that life is a physicochemical phenomenon that can happen on other planets. Today those ideas may seem logical, but more than 100 years ago he was certainly ahead of his time.

2.1.2 Origins in the Lab

Alicia Negrón liked science, she does not remember why; her family was into politics, so she did not have role models, except Marie Curie. Chemistry was her passion despite the bad teachers during most of her basic education. A high school teacher changed the pattern and convinced Alicia that organic chemistry was what she wanted to study. She graduated as chemical pharmacobiologist at the Universidad Nacional Autónoma de México (UNAM) in 1968; by then, she was interested in the chemical evolution of life and decided to study a PhD in the laboratory of Cyril Ponnamperuma, an expert of the chemical evolution of life, that was involved with NASA in the Viking and Voyager programs and was one of the founders of the International Society for the Study of the Origin of Life (ISSOL) that became The International Astrobiology Society in 2005. In his laboratory, they analyzed organics from meteorites and lunar rocks brought by the Apollo missions (Navarro-González 1998). In 1976, Alicia got a job at the UNAM, where she started the Laboratory of Chemical Evolution.

Alicia taught at the school of chemistry (Facultad de Química) and the school of sciences (Facultad de Ciencias); she was a very good teacher and students soon arrived at the new laboratory. Over time, the laboratory grew with more instruments, space, two colleagues, and students. Nowadays, there are between 15 and 20 students at any time, working at her laboratory from undergraduate to graduate programs of chemistry, biology, physics, and Earth sciences. In her laboratory, Dr. Negrón explores the role of high-energy radiation like gamma ray and particles like cosmic rays in the chemical evolution of life. She has proposed that clays are key on the formation and organization of the molecules that led to first cells. One of the many students who arrived at the Laboratory of Chemical Evolution was Rafael Navarro-González.

Rafael was a bachelor student when Cyril Ponnamperuma visited the UNAM. They had a casual talk after Ponnamperuma's conference, and he invited the student to come to Maryland to study in the chemistry graduate program after he finished his undergraduate studies under the supervision of Alicia Negrón. Then he fulfilled his goal of earning a PhD with Ponnamperuma as his advisor in 1989. He returned to UNAM in 1994 to the Laboratory of Plasma Chemistry and Planetary Studies, which had just started operations. In 1995, Mexico had the first student to earn a degree with a thesis focused in astrobiology, Alfredo Romero Delgado from the biology bachelor program at UNAM. After Alfredo, many other students found in Navarro's laboratory the place to start their career in astrobiology.

2.1.3 The Mexican Society of Astrobiology

For most authors on this chapter (as being part of the Society's Executive Councils), keeping alive a Society dedicated to boost astrobiology in Mexico has been a huge lifetime challenge. The Sociedad Mexicana de Astrobiología (SOMA, www.soma. org.mx) is a nonprofit, self-sustained organization comprised of a multidisciplinary group of researchers, teachers, students, and anyone else within the general public interested in supporting the development of astrobiology in Mexico. Its headquarters were initially located at Instituto de Ciencias Nucleares at the UNAM campus in Mexico City. SOMA's roots date back to November 2000, when the Sociedad Mexicana de Ciencias de la Vida en el Espacio (Mexican Society of the Sciences of Life in Space) was created by Mexican researchers and people dedicated to the popularization of science. By February 2002, the Society adopted its current name and identity. Without a doubt, SOMA can be considered the major self-organized assembly of astrobiologists in Mexico, not only it harbors between 50 and 100 members each year from most country states, but it has also established partnership agreements with the NASA Astrobiology Institute and the European Astrobiology Network Association since 2012.

During its almost 20 years of existence, SOMA has managed to hold 11 National Meetings, five sessions of the Mexican School of Astrobiology, nine research seminars to celebrate the "Day of Astrobiology," and multiple science outreach events every year.[1] With the exception of the National Meeting, SOMA's activities are free of charge. The National Meeting and the Mexican School of Astrobiology (EMA) represent two of SOMA's major academic activities, taking place every 2 years to promote national and international research collaborations, as well as transdisciplinary training and discussions, on a wide range of subjects related to Astrobiology. We are particularly proud of three projects. First, a new strategy for the last EMA – held last September 2019 in Puebla at the National Institute of Astrophysics, Optics and Electronics (INAOE) – was enforced toward teaching the scientific background and research being developed for two complementary topics in astrobiology (e.g., exoplanets and extremophiles), through an intense schedule of courses, research talks, outreach activities, training on field sites, computational analysis, laboratory experiments, and telescope observation. Second, a collective effort of 10 years will conclude on 2020 with the publication of the first academic book on astrobiology in Spanish for Latin America, entitled "Astrobiología: una visión transdisciplinaria de la vida en el Universo," in a co-edition between UNAM and Fondo de Cultura Económica. Lastly, we are moving forward to actively engage Mexican astrobiologists on research collaborations by documenting and providing information about field sites of astrobiological interest.

---

[1] SOMA activities are supported by SOMA members, public Mexican Universities, and other national and international institutions. Most funds are provided by the Mexican National Council of Science and Technology (CONACYT).

2.1.4   Field Sites of Astrobiological Research Interest in Mexico

On her own words, Janet Louise Siefert – an Associate Research Professor at Rice University and the NASA Astrobiology Institute – became an astrobiologist after 10 years of being a full-time real housewife (Siefert 2011). Along with Valeria Souza and Luis Eguiarte at UNAM, she has dedicated 20 years of research to understand the biological diversity and ecological structure of the Cuatro Ciénegas Basin (CCB), Coahuila, one of the most enigmatic Mexican field sites (Souza et al. 2012). Mexico is within the world's top five mega diverse countries, so the CCB is just one of the almost 30 field sites of astrobiological interest that SOMA has been able to empirically track throughout the country. These field sites are extreme environments located throughout a large diversity of natural settings, such as lakes, ponds, lagoons, basins, oceanic hydrothermal vents, geothermal fields, caves, volcanoes, and phreatic sinkholes (cenotes). Some of them are shown in Fig. 2.1. Unfortunately, less than a third of such field sites has been studied for a full microbial characterization. Moreover, some of the developed studies do not include the expertise of Mexican researchers, although others have been cleverly performed by Mexican scientists with international collaboration of well-known astrobiologists, like Janet Siefert, Christopher P. McKay (NASA Ames Research Center), and Purificación López-García (French National Centre for Scientific Research).

Astrobiologists develop diverse interests by exploring the geochemical profiles and characterizing the phylogenetic types (phylotypes) of these field sites, mainly represented by members of the "rare biosphere" (see Table 2.1). But in general, we all are searching for any signal or process in the environment pointing out to a self-sustained disequilibrium in the geochemical cycles. After careful examination, such disequilibrium might indicate that life is or was present on such environment. For instance, some extreme environments are considered analogous metabolic ecosystems for Mars, Europa (one of the Jupiter's moons), Enceladus (one of Saturn's moons), or Earth's earliest life. Other environments instead might host extremophiles or biosignatures that can help us to unambiguously detect past or present life on another world (see Chap. 1). Roughly speaking, a biosignature is any detectable substance (e.g., element, isotope, molecule) or phenomenon that provides scientific evidence of extant or ancient life. On the other hand, extremophiles are organisms able to metabolically operate under physical (e.g., pressure, temperature, radiation) or geochemical (e.g., pH, salinity, desiccation) extremes, and whose adaptive responses and survival thresholds have evolved to grow optimally under one or more of these extreme conditions. Hydrothermal vents, microbial mats, and microbialites embrace most astrobiological interests: analogous terrestrial metabolisms, biosignatures, and extremophiles. In contrast to hydrothermal vents (Martin et al. 2008), microbial mats and microbialites occur in several environments throughout Mexico (Beltrán et al. 2012). Some examples are shown on Fig. 2.1 and are also described in detail throughout the book. Cuatro Ciénegas is in the Chihuahua Desert. The basin is surrounded to the northwest by the mountain range Sierra La Madera,

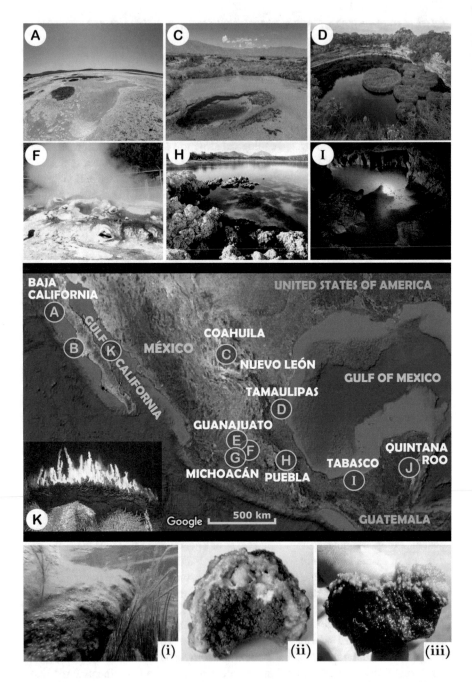

Fig. 2.1 Examples of field sites of astrobiological research interest in Mexico. CENTRAL PANEL shows the geographical location of the field sites on the satellite map of Mexico (retrieved from Google Maps 2019, with information of NASA, NOAA, and INEGI). UPPER PANEL shows the landscape views of some of the field sites explored by astrobiologists and students from the

to the west by Sierra de la Fragua, to the south by Sierra de San Marcos, and to the north and northeast by Sierra de Alamillos and de la Fragua. These mountains protect the ponds in Cuatro Ciénegas. The mountain ranges (sierras in Spanish) retain the oasis humidity and prevent from light pollution generated by the municipality of Monclova (Falchi et al. 2016). Cuatro Ciénegas night sky is illuminated only by the Milky Way, the Moon, and sometimes by a shooting star, a bolide, or a superbolide. Next, given its astrobiological importance, we will review the difference between these phenomena, their relationship with the objects that create them, and their origin. Afterwards, we will discuss some of the history of the meteorites that fell in Coahuila and the importance of protecting them as a national heritage.

2.2   The Rocks That Fell from the Sky

In the context of the class Geology and Planetary Atmospheres taught at the School of Science (UNAM), the first subject that we teach is the difference between the terms: asteroid, meteoroid, micrometeoroids, meteor, and meteorite. According to Rubin and Grossman (2010) given the size of the material, micrometeoroids size is between 10 µm and 2 mm, meteoroids is between 2 mm and 1 m, while asteroids and comets are bodies of more than 1 m; all these objects are currently in orbit. The term meteor refers to the phenomena that occur in the atmosphere, when a solid body enters the planetary atmosphere at speeds larger than 7 km/s and the atmosphere exerts resistance-causing fusion of the material and even partial ionization. When a micrometeoroid enters the atmosphere, it creates a narrow light track called shooting star. When a meteoroid enters the atmosphere, it generates a ball of fire or bolide brighter than the planet Venus. But when an asteroid enters the atmosphere, a superbolide is generated, accompanied by fragmentation of the material and shockwaves, and if the asteroid has a diameter larger than 50 m, it creates an impact crater on the surface. Micrometeorites (10 µm – 2 mm) and meteorites (>2 mm) are solid objects originated in a celestial body but that are found on the surface of planet

◄──────────────────────────────────

Fig. 2.1 (continued) Mexican School of Astrobiology (EMA). Locations are: (**A**) Figueroa Lagoon, Baja California (Photo credit: Paco Beretta, III EMA 2015); (**B**) Guerrero Negro salt flats, Baja California; (**C**) Pozas Azules, Cuatro Ciénegas Basin, Coahuila (Photo credit: National Park Service, USA, nps.gov); (**D**) Cenote "El Zacatón," Tamaulipas (Photo credit: Art Palmer, from astrobio.net); (**E**) Crater-lake Rincón de Parangueo, Guanajuato (see photo on Fig. 2.4); (**F**) Los Azufres, Michoacán (Photo credit: Raúl López Mendoza from cambiodemichoacan.com.mx); (**G**) Paricutín and Sapichu volcanoes, Michoacán; (**H**) Alchichica crater-lake, Puebla (Photo credit: students from IV EMA 2019); (**I**) "Luna Azufre" Cave, Tabasco (Photo credit: Michael Tobler); (**J**) Bacalar Lagoon, Quintana Roo (see photo on (i)); (**K**) Pescadero Basin, Gulf of California (Photo credit: Ronald Spelz from MBARI, USA, mbari.org). DOWN PANEL shows cross-sections of extant microbialites and biomats. (*i*) oxygen being released by stromatolites under water (seen as bubbles over the green biomat) at Bacalar Lagoon (Photo credit: Kamila Chomicz from nytimes. com). (*ii*) spongy microbialites from Alchichica crater lake (Photo credit: Beltrán et al. 2012, Aquat. Microb. Ecol.). (*iii*) hypersaline microbial mat at Figueroa Lagoon, different layers of colors are made up by different phylotypes. (Photo credit: Ruth Quispe, III EMA 2015)

Table 2.1 Description of some of the most studied field sites with astrobiological research interest in Mexico

Site	Site coordinates and description	Microorganisms detected	Astrobiological interests
(A) Figueroa Lagoon, Ensenada, Baja California (a.k.a. Mormona Lagoon) (1)	30°38′41.3″ N; 116°01′08.3″ W A closed hypersaline lagoon, almost completely filled with silica particles and gypsum. The main source of water is the ocean, seeping through the dunes at high tide and forming permanent puddles or ephemeral pools (with meters or kilometers in diameter)	Thick, laminated, endolithic, and hypersaline biofilms comprised predominantly of oxygenic and anoxygenic phototrophs, sulfate- and methane-reducing bacteria. Cyanobacteria such as Microcoleus are living in water 2–3 times as salty as seawater. See references for a full list of found phylotypes	Mars: A model to redefine the isotopic boundaries of biogenic methane (a biosignature), helpful to test for false negatives returned from measurements of methane on Mars and other planetary bodies
(B) Salt flats at Guerrero Negro, Baja California (1)	28°06′16″ N; 114°09′10″ O A 250 m^2 area of shallow (~1 m) evaporating basins, with high salinity (50–130 ppt). Gypsum deposits are formed in saltern seawater concentration ponds and in columnar microbial mats located on natural anchialine pools		
(K) Guaymas Basin, Gulf of California (2)	23°38.5′ N; 108°23.6′ W A system of hydrothermally active environments (e.g., Pescadero and Alarcón Rise vents) that includes vent plumes, seeps, and anoxic sediments, each of them exhibiting a wide range of pH and temperatures	Surface-attached microbial mats, diverse prokaryotic and eukaryotic microbes, and symbiont-harboring invertebrates See references for a full list of found phylotypes	Europa: A model system to study anaerobic microbial communities (methane oxidizers, and sulfate reducers) that catalyze a sulfate- and methane-driven anaerobic carbon cycle and that is compatible with stable carbon and sulfur isotope evidence for early Earth ecosystems
(C) Cuatro Ciénegas Basin (CCB), Chihuahuan Desert, Coahuila (3)	26°50′53.19″ N; 102°08′29.98″ W A large number of highly diverse springs (>300, e.g., Pozas Azules), spring-fed streams, and terminal evaporitic ponds form an inverse archipelago, in which aquatic systems are separated by sparse desert vegetation, microbial crusts, and salty soils. It is a phosphorus-limited environment	One of the most microbial diverse sites on Earth for viruses, bacteria, and archaea, including several ancient species and endemic species (e.g., 220 new *Bacillus* linages). See references for a full list of found phylotypes	Mars: A proxy for the late Precambrian, when eukaryotic organisms diversify and became more complex. It was suggested as the most analogous target to Gale crater from the MSL/Mars 2018 mission

(continued)

Table 2.1 (continued)

Site	Site coordinates and description	Microorganisms detected	Astrobiological interests
(D) Cenote "El Zacatón," Tamaulipas (4)	22°59′35.10″ N, 98°09′55.96″ W The world's deepest limestone, vertical phreatic sinkhole (cenote). It has 105 m in diameter, 318 m deep, and a 17 m deep limestone rim/cliff band above the water; elemental sulfur clouds develop in sunlight. Its hydrothermal activity has considerably calmed down, but there are signs of an unusually well-mixed environment: a pervasive scent of sulfur hovers around the cenote, and its water has a constant temperature of 30 °C	Low diversity in water column (1–273 m): mostly photosynthetic anoxygenic sulfide oxidizers Gamma- and Epsilon-proteobacteria, Chlorobi; Crenarchaeota. High diversity in wall biomats (1–273 m): chemoautotrophic anoxygenic sulfide oxidizers Chlorobi, Chloroflexi, and Deltaproteobacteria; Crenarchaeota, Methanomicrobia, anaerobic methane oxidizers (ANME-1), and the candidate phylum Azufrosa	Europa: A model for chemoautotrophic, anaerobic sulfide oxidation (independent of photosynthetically derived carbon) in deep, dark, extraterrestrial aqueous systems with a hydrothermal feed source. It is investigated by the Deep Phreatic Thermal Explorer (DEPTHX), an autonomous robotic underwater vehicle designed to explore signs of life in the deep ocean under the icy crust of Europa
(E) Crater-lake Rincón de Parangueo, Guanajuato (5)	20°25′53.4″ N; 101°15′00.7″ W A Quaternary maar (<0.137 Ma) located at the central part of the TMBV. A perennial lake was present until the 1980s. Nowadays, most of the lacustrine area has been transformed in a subaerial alkaline soil surface, but since 2009, there is always at least a remnant hypersaline and alkaline (pH 9–10) pond inside the crater	Hygroscopic cyanobacteria in the non-lithified crustal layers of individual thrombolites. Microbialites and endolithic microbial mats in hypersaline ponds, including phylotypes from bacteria (Cyanobacteria, Proteobacteria), Euryarchaeota, and Eukarya (Chlorophyta, Bacillariophyta, Rotifera)	Mars and Enceladus: Analogous microbial ecosystems for extant and ancient life. Characterization of extremophiles and biosignatures

(continued)

Table 2.1 (continued)

Site	Site coordinates and description	Microorganisms detected	Astrobiological interests
(F) Los Azufres, Michoacán (6)	19°46′51.7″ N; 100°39′23.6″ W A geothermal field located within the silicic volcanic complex in the TMVB (3000 m a.s.l.). It comprises many natural hydrothermal springs, fumaroles, and boiling mud pools (pH 5.5–7.4, 25–300 °C). It has the highest natural arsenic concentrations in America (24 mg/L)	Mesophilic, hyperthermophilic and acidophilic microorganisms. Anoxygenic phototrophs in biomats (Rhodobacter), colorless sulfur oxidizers in mud and water (*Acidithiobacillus, Lysobacter*); sulfate and sulfur reducers (*Thermodesulfobium, Desulfurella*); *Acidithiobacillus, Thiomonas*	Analogous microbial ecosystems for extant and ancient life on other worlds. Characterization of extremophiles and biosignatures
(G) Paricutín and Sapichu volcanoes, Michoacán (7)	19°29′36.8″ N; 102°15′04.4″ W The Paricutín is the most recently active monogenetic cinder cone in the MGVF (at 2808 m a.s.l.). It has vapor steam vents that are slightly acid to neutral (pH 5.3–7.9), with different temperatures (soil: 27–87 °C; air: 13.6–56 °C), and high concentrations of metals such as iron and arsenic	Thin biomats over mineral crusts: photosynthetic bacteria (Cyanobacteria, Chloroflexia class), Actinobacteria, Proteobacteria, Ktedonobacteria, Acidobacteria; anaerobic bacteria: Clostridia, Bacilli. Suggested metabolisms include sulfur respiration, nitrogen fixation, methanogenesis, carbon fixation, photosynthesis	Analogous microbial ecosystems for extant and ancient life on other worlds. Characterization of extremophiles and biosignatures
(H) Alchichica Lake, Puebla (8)	19°24′51.5″ N; 97°24′10.9″ W A soda, athalassohaline crater-lake located at 2300 m a.s.l., and a depth of 64 m. It is a hyposaline (8.5–10 g/liter), alkaline (pH 8.9–9.3), and volcanic monomictic lake, i.e., stratified during most of the year, the oxygenated surface water mixing with deep anoxic water only occurs during the winter season. Its water is oversaturated with magnesium and calcium carbonates (Mg:Ca = 40): $Na^+ > Mg^{2+} > K^+ > Ca^{2+}$	Living columnar and spongy microbialites down to at least 14 m deep. Dominant phylotypes are oxygenic (cyanobacteria, green algae, and diatoms) and anoxygenic phototrophs (Alphaproteobacteria, Chloroflexi); heterotrophic bacteria (sulfate reducers) and eukaryotes (fungi and a huge diversity of protists); Crenarchaeota	Mars: Microbial ecosystems analogs for extant and ancient life on Mars. Identification and characterization of biosignatures that can be used to detect extant life in Martian carbonates

(continued)

Table 2.1 (continued)

Site	Site coordinates and description	Microorganisms detected	Astrobiological interests
(I) "Del Azufre" and "Luna Azufre" Caves, Tabasco (9)	17°27′09.8″ N; 92°46′41.6″ W These are mostly permanent dark caves. Inside the "Del Azufre" cave, several springs discharge water that is rich in hydrogen sulfide (H$_2$S). Its water sulfide is oxidized to colloidal sulfur, which gives the water a milky appearance. "Luna Azufre" is a non-sulfidic cave	Chemoautotrophic bacteria in snottites. A particular dominant phylotype is sulfur-oxidizing *Acidithiobacillus* spp. (pH 0–1.5)	Analogous microbial ecosystems for extant and ancient life on other worlds. Characterization of extremophiles and biosignatures
(J) Bacalar Lagoon, Quintana Roo (a.k.a. "Lagoon of the seven colors") (10)	18°46′03.9″ N; 88°18′24.7″ W An oligosaline karstic coastal, freshwater system (15 m deep, pH 7.8, 25–28 °C) with the largest (10 km) and oldest (from early Holocene) occurrence of extant microbialites. Ca^{2+} cation concentrations are close to marine values (Mg:Ca = 0.24), and HCO$_3$ values exceed average marine concentrations: Ca^{2+} > Mg^{2+} > Na$^+$	Microbialites include domes, ledges, and oncolites and exhibit a high content of organic carbon, as well as C:N, C:P, N:P ratios. It has a high genetic diversity (equally to that found in CCB). Dominant phylotypes are Cyanobacteria, Proteobacteria, and Crenarchaeota. See references for a full list of found phylotypes	Analogous microbial ecosystems for extant and ancient life on other worlds. Characterization of extremophiles and biosignatures

(1) Ley et al. (2006); Vogel et al. (2009); Potter et al. (2009); Tazaz et al. (2013); Harris et al. (2012); Valdivieso-Ojeda et al. (2014). Main Mexican PIs: Jacob A. Valdivieso-Ojeda, Miguel A. Huerta Diaz, Francisco Delgadillo-Hinojosa
(2) Teske et al. (2003); Goffredi et al. (2017). Main Mexican PIs: Elva Escobar Briones, Ronald M. Spelz Madero
(3) Souza et al. (2012); Centeno et al. (2012); Tapia-Torres et al. (2016); De Anda et al. (2018); Valdespino-Castillo et al. (2018). Main Mexican PIs: Valeria Souza Saldivar, Luis Eguiarte Fruns, Gabriela Olmedo Álvarez, Luisa I. Falcón
(4) Krajick (2007); Sahl et al. (2010)
(5) Montoya et al. (2017); Chacón et al. (2018); Sánchez-Sánchez et al. (2019). Main Mexican PIs: Elizabeth B. Chacón, Lilia Montoya Lorenzana, Luis M. Cerca Martínez
(6) Brito et al. (2014); Chen et al. (2018). Main Mexican PIs: Elcia M. Souza Brito, Luis E. Servín Garcidueñas
(7) Medrano-Santillana et al. (2017); Brito et al. (2019). Main Mexican PIs: Elcia M. Souza Brito
(8) Navarro et al. (2010); Couradeau et al. (2011); Beltrán et al. (2012); Valdespino-Castillo et al. (2018). Main Mexican PIs: Elva Escobar Briones, Javier Alcocer Durand, Luisa I. Falcón, Rosaluz Tavera Sierra
(9) Tobler et al. (2008); Jones et al. (2016)
(10) Gischler et al. (2008); Centeno et al. (2012); Valdespino-Castillo et al. (2018). Main Mexican PIs: Luisa I. Falcón
The capital letters within parentheses (depicting the name of each field site) correspond to the ones showing their geographical location on the map and landscape views on Fig. 2.1, numbers in parentheses are for references, and PIs listed in the Table footnote. Abbreviations, *PIs* principal investigators, *a.s.l.* above the sea level, *ppt* parts per thousand, *TMBV* Trans-Mexican Volcanic Belt, *MGVF* Michoacán-Guanajuato Volcanic Field

Earth. In the following section, we present the general classification of meteorites and the importance of reporting if one is found or if a fall is witnessed.

2.2.1 The History of the Solar System Written in Rocks

Some meteorites are shiny and heavy, others dark and light, some others are difficult to distinguish from a rock. Each of those characteristics is the result of processes that happen before, during, and after the formation of the planets of the Solar System. Classifications are the first step to understand those processes. For meteorites, one type of classification is based on the processes they have experienced, leading to two main groups, undifferentiated and differentiated.

Undifferentiated meteorites are chondrites, which at some point were part of parent bodies that were not affected by processes of fusion and differentiation, evident by the uniform and fine dispersion of the metal (iron and nickel alloys) and iron sulfide (troilite) grains (Fig. 2.2). The most characteristic components of these meteorites are chondrules. If a full analysis of each of the inner planets (Mercury, Venus, Earth, and Mars) were done, their composition would be essentially chondritic; thus, chondrites contain all the elements present in the inner planets. The carbonaceous meteorites are those that contain the highest quantity of hydrated minerals and the abiotic organic compounds. The most ancient components of the Solar System are calcium and aluminum inclusions (CAIs), which are dated to 4567 million years (Myr) and are contained in this type of meteorites. Given the accretion process, the chondrules and the calcium and aluminum inclusions

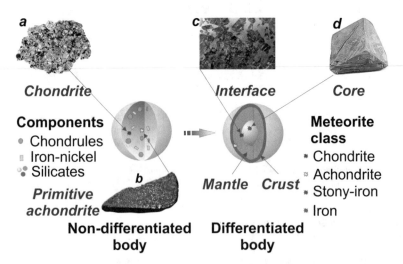

Fig. 2.2 Classification of meteorites according to the differentiation of their parent body. Examples of meteorites are (**a**) Allende chondrite, (**b**) Acapulco primitive achondrite, (**c**) Brenham stony-iron, (**d**) Gibeon iron. (Photos of meteorites: Martha Garay and Ana Emilia Pérez)

agglomerate with small olivine and pyroxene crystals to form bodies that grew to form rocks, planetesimals, and planetary embryos until they formed the planets as we know them. Planetesimals or part of planetary embryos that fragmented are represented by the asteroids and fragments of comets that we know today. The source of chondrites are the asteroids, although some authors propose that comets are the source of some carbon chondrites and micrometeorites.

Differentiated meteorites include the achondrites, the mixed meteorites, and the metallic meteorites, all of them come from bodies that were large enough for their self-gravity to induce a chemical differentiation (Fig. 2.2). The complete fusion of the interior caused the separation of the silicate and metal liquids, each component with a distinct density, in such a way that the light silicate liquid surrounds a metallic core of iron and nickel that concentrates in the center of the body because it is the heaviest, and then, they cooled passing from liquids to solids. Silicates are compounds of silicon, oxygen, and cations such as iron, magnesium, calcium, sodium, and potassium. The crust and mantle of Earth are made of silicates.

The cooling of these bodies was followed by collisions in space. Thus, the fragments of the core are now what we know as metallic meteorites; the mixed meteorites were part of the transition zone between the metallic core and the silicate mantle; finally, the achondrites are the fragments of what was the mantle and crust of the parent body. The differentiated meteorites support the theory that the interior of the rocky planets are also separated in a metallic core in the center of the planet, surrounded by a mantle and covered by a silicate crust. The source of these meteorites is some asteroids and certain planetary surfaces such as Vesta, the Moon, and Mars.

Have meteors been sighted close to Cuatro Ciénegas? Recently, there have been no reports of fallen material, but the Meteoritical Bulletin Database reports five meteorites found in the entire state of Coahuila. These are an ordinary chondrite named La Esmeralda and five metallic meteorites named El Burro, Coahuila, Coahuila II, Puente de Zacate, and Acuña.

2.2.2 From the Sky to Mexico: Mexican Meteorites and Their Study

In Mexico, meteorites are being studied in three different epochs marked by the publication of different catalogs. Antonio del Castillo, an engineering geologist, established in Mexico the study of meteoritics since 1864 with the description of the iron meteorite Yanhuitlán up to the publication of the catalog (1889) and a meteorite distribution map (1893). When the first catalog had been finished, the engineering geologist José Haro published, in 1931 a second catalog.

Meteorites in Mexico were studied outside of the territory during the period encompassing 1930 to 1978. In 1969, in Pueblito de Allende, Chihuahua (320 km to the west of Cuatro Ciénegas), an event that revolutionized the meteoritic science at a global level occurred. Between 0105 and 0110 Central Standard Time, people located in the South of the United States of America (in Nuevo Mexico and Texas) observed the entry of a bright bolide, while the inhabitants of Parral, Chihuahua, heard a very strong noise followed by a strong air blast (King et al. 1969). The explosion created thousands of fragments that were dispersed over an ellipse of more than 50 km in elongation. The original rock is a carbonaceous chondrite CV3 (Fig. 2.2a) and by that time the knowledge about the fragments was limited. The Smithsonian Institution collected more than two tons of material. The smithsonian was able to distribute four kilograms of pulverized material to the laboratories that would later receive the lunar samples (Mason 1975). As chondrites are the precursor material of rocky planets, the evolution of the planetary bodies is studied comparing the differences in the chemistry of the rocks from a planetary body with respect to the initial chondritic material. The scientist found calcium aluminum inclusions, a high-temperature mixture of minerals that are more than 4567 million years old; it means the beginning of the Solar System (Connelly et al. 2012). The study of Allende drove the start of cosmochemistry, at the intersection of geology, planetary science, astronomy, and astrophysics. A big piece of the Allende rock is in display in the Museo de Geología in downtown Mexico City (Mason 1975).

In August 11, 1976, at 11:00 a.m. local time, near to the Acapulco beach, Leodegario Cardenas heard an airplane-like sound, and when he turned-up his face to the sky, he saw a bolide falling to him. Fortunately, the rock fell near to his farmed land and he picked-up a 10-cubic centimeters blacked sample. A few years later, he visited Arcadio Poveda, a prominent Mexican astronomer, and the geologist Gerardo Sánchez Rubio. Sánchez Rubio reported the event to the Scientific Event Alert Network (SEAN 1978) and to the Meteoritical Bulletin (Sánchez Rubio 1978). This rock has a chondritic composition, but chondrules are not present (Fig. 2.2b). The rock is a primitive achondrite, and it was a missing link between chondrite parental bodies and the evolved planet, that way a new group of meteorites was born: the acapulcoite primitive achondrite group (Palme et al. 1981).

In 2013, the teacher Elvia Muñiz took a rock to the Museum of Geology because she thought it could be a meteorite. After a special procedure, the Widmanstatten structures appeared, these structures are made by two minerals of iron and nickel (taenite and kamacite) that seem interlaced like the threads in an artisan fabric (this pattern is shown in Fig. 2.2d). This 2.3 kg rock was an iron meteorite. It was registered as the Mexican meteorite number 100 (Cervantes et al. 2017). The study of this piece of cosmic material allowed, for the first time, in Mexico, to determine abundances of some elements such as Ge, Ga, Mo, Ir, Pd, Sn, Co, CR, Pt, In, and Re. This meteorite is a medium octahedrite called Tequisquiapan, and its name is due to the nearest principal town, located in Queretaro State, in Central Mexico.

2.2.3 Understanding Alien Rocks: Research in México

On October 4, 1990, a group of researchers of the UNAM participated in the Colloquium of the Meteorites (Ruvinovich-Kogan 1992), during which planetary science subjects, such as impacts, meteoritics, and planetary geology, were discussed. Since then, the study of meteoritics in Mexico has grown, for example, between 2003 and 2019, and 21 students have graduated with thesis about meteoritics. In 2001, Sánchez Rubio and collaborators published a catalog of the meteorites of Mexico in which they report Allende CV3 and Acapulco (acapulcoite). In this catalog, the present location of most of the Mexican meteorites is reported, in particular at the collections of Instituto de Geología and Instituto de Astronomía, both at UNAM.

In 2003, a meteoritics group was created and led by researcher Fernando Ortega Gutiérrez to study mainly the ages of some Mexican meteorites (Hernández-Bernal and Solé 2010), the possible sources of chondritic meteorites (García-Martínez and Ortega-Gutiérrez 2008) and the formation and thermic evolution of minerals in ordinary chondrites (Cervantes-de la Cruz et al. 2010; Alba-Aldave et al. 2009). The group also studies the formation of structures within meteorites, such as chondrules (Cervantes-de la Cruz et al. 2015) and chondrites (Reyes-Salas et al. 2010; Corona-Chávez et al. 2018). Magnetism of planetary bodies that is found as a remanent in meteorites has also been studied in Mexico (Flores-Gutiérrez et al. 2010a, b; Urrutia-Fucugauchi et al. 2014).

2.2.3.1 A Multidisciplinary Group to Study Meteorites

Geologists study rocks and astronomers study stars, their expertise seem very far away from each other, but for meteoritics, we need both. Geologists can tell what a rock is made of, the processes it has been through, and its origin. Astronomers are in charge of understanding how planets form, in particular those in our Solar System. Meteorites formed during the first stages of our planetary system and are our best samples available on Earth to understand what happened. In 2013, Karina Cervantes de la Cruz, a geologist and Antígona Segura, an astronomer and both authors of this chapter, formed an interdisciplinary group to study the evolution of minerals in meteorites, since their formation until the years that they spend on Earth. The group included two experimental physicists dedicated to quantum optics and undergraduate and graduate students of Earth sciences, engineering geology, and astrophysics. The group has dedicated to:

1. Study the evolution of minerals in chondritic meteorites from their formation and their transformation due to heating processes in the parental bodies. These heating processes are due to the radioactive decay of elements such as ^{26}Al. Heat transforms minerals from one crystalline form to another, and the process occurs from the edges of the mineral toward the center (Cervantes-de la Cruz et al. 2010, Alba-Aldave et al. 2009).
2. Understand the origin of chondrules. Chondrules are spheres of micrometer to millimeter sizes. They are composed of olivine and pyroxene, which are silicate minerals rich in iron, magnesium, and calcium (Montoya-Pérez 2016, 2019). Their spherical form and igneous texture suggest that the minerals were melted at some point and that they cooled down very quickly. Scientists have not agreed

on how chondrules were formed. Our approach is to create chondrules in the laboratory by melting silicates with an experimental device designed and created for such purpose called Citlalmitl (Cervantes-de la Cruz et al. 2015).

3. Study the terrestrial weathering and the degree of deterioration due to atmospheric and climatic processes in iron meteorites (González-Medina 2019, López-García 2019).

2.3   Planetary Collisions

Six planetary processes have been recognized on planetary bodies with solid surfaces: tectonics, volcanism, impact cratering, weathering, mass wasting, and interaction with surface fluids (this last one includes eolian, fluvial, and glacial processes) (Melosh 2011). Not all of these processes have occurred on planetary surfaces, and some of them occurred in the past but not currently. For example, on Mars, water flowed on its surface, but this process ended in the Noachian, a geological Martian period that lasted between 4100 and 3700 billion years ago.

Among the previous processes, the only one that has happened along all the history of our planetary system on all (or almost all) planetary surfaces is impact cratering. This is evident from the heavy cratered surfaces of Mars, Mercury, the Moon, Calisto, and even asteroids. Counts of impact craters on the Moon's surface have shown that from 4500 to 3800 billions of years ago, there was a heavy impact bombardment of objects, and that from ~3800 billions of years ago to the present time, the rate of impacts has been more or less the same, even though some authors think that the cratering rate in the inner Solar System decreased by a factor of two to three during the past 3 billions years, but it increased during the last 500 millions years, probably due to the catastrophic disruption of a large main-belt asteroid (Michel and Morbidelli 2013).

Studying impact craters has been very useful to understand rheology of planetary crusts, assess relative and absolute ages (and as a consequence to know about the geological histories of planets and moons), and recognize the role of impacts on the life on Earth (Osinski and Pierazzo 2013). When an asteroid or a cometary nucleus impacts on a planetary body, different situations can happen. Impacts on airless bodies produce impact craters from microns to hundreds or thousands of kilometers, depending on the impactor kinetic energy. On the other hand, if the target has an atmosphere, then the impactor interacts with it and a meteor or a fireball is produced. Due to the drag pressure that the atmosphere exerts on the impactor, it can be fragmented into pieces, each one of them produce meteors and can fragment again (Hills and Goda 1993; Ceplecha et al. 1993). If the impactor (or its fragments) has enough kinetic energy, its impact on planetary surfaces produces impact craters (Passey and Melosh 1980), otherwise if the original body or its fragments survive their passage through the atmosphere and we can collect them, they are called meteorites.

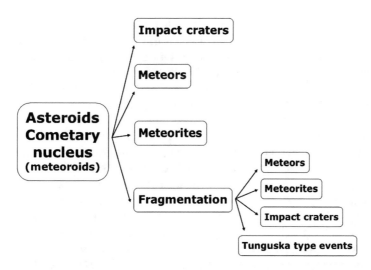

Fig. 2.3 Possible situations when asteroids or cometary nucleus impact planetary surfaces

On the morning of June 30, 1908, an object of several tens of meters long burst over the Siberian taiga at a height between 6 and 10.5 km, with an energy between 10 and 15 megatones. The object did not touch the surface, but the shock wave destroyed 2150 ± 50 km^2 of taiga (Farinella et al. 2001). As far as we know, this is the only event of this type registered until now, but there could be more in the future. Fig. 2.3 shows a diagram that summarizes the phenomena that can happen due to impacts of asteroids or cometary nucleus with planetary surfaces.

2.3.1 Impact Studies from Mexico

At UNAM, we have studied several of the previous phenomena, in particular, impact craters, meteorites, and atmospheric bursts. In the next paragraphs, we briefly describe some of this work.

Examining the frequency distribution of diameters of the fragments produced by an impact by Dohnanyi's distribution (Dohnanyi 1969), the relation between the maximum diameter of the fragments ejected from the primary craters and their ejecta velocities done by Vickery (1987), and the total mass of ejecta with escape velocity due to an impact (O'Keefe and Ahrens 1977; Melosh 1989), Cordero Tercero and Poveda calculated the number of escaped fragments larger than 2 cm and 10 cm as a result of the excavation of Chicxulub impact crater. They compared the ratio of the number of Chicxulub ejecta with a specific size to the number of NEOS of the same diameter and concluded that it is possible that some fragments of Chicxulub ejecta entered to Earth atmosphere after circulating the inner Solar System for some time (Poveda and Cordero 2008). The falling frequency of this material was larger in the past than nowadays. It is possible to find Chicxulub meteorites on the Moon and even on Earth, but in the latter case, it would be more difficult to recognize them. This one was one of the first works done on this topic, and their authors realized how interesting it is, as well as the possibility to understand interactions between cosmic bodies (Poveda and Cordero 2008).

In 1931, the priest Fedele da Alviano reported that in the morning of August 13, 1930, people that lived near Curuça River, near the border between Brazil and Peru, witnessed the fall of three objects together with the falling of dust, earth tremors, and terrible sounds (L'Osservatore Romano 1931). This news was known by Leonid Kulik who wrote a short note about it and called it "The Brazilian Tunguska" (Kulik 1931). Fifty-eight years later, Vasilyev and Andreev (1989) made some comments to Kulik's note and wrote a kind of invitation to South American scientists to study this event. Bailey et al. (1995) analyzed the three previous papers, suggested that this event could be produced by a fragment of Perseids, and reiterated the invitation to South American colleges. Finally, Reza and colleges studied Curuça event using remote sensing images, seismic records, and fieldwork (Reza et al. 2004). After analyzing these works, Cordero and Poveda started a Sherlock Holmes style research in which they study each line of evidence done by the original Da Alviano's report and the Bailey's translation (Cordero and Poveda 2011). In this work, they used information about nuclear weapons, scaling laws, empirical data, and numerical simulations to study several topics related to Curuça event: a possible impact crater related to this event, height of explosion, energy of explosion, seismic records, sounds, and its possible relation with the Perseids meteor shower. Their conclusions were that neither the circular feature found by Reza and coauthors nor the seismic record obtained by them from the San Calixto seismic observatory in La Paz, Bolivia, are related to Curuça event. Moreover, they have evidences (not published yet) that the mentioned circular feature is not an impact crater. Other conclusion was that the object that produced the Curuça event could be a 9 m long object that exploded at a height of 6 km releasing an energy less than 9 kilotons of TNT and entered the Earth's atmosphere at 11.2 km s^{-1}. So, this event was three orders of magnitude less energetic than Tunguskas's.

On February 10, 2010, around 15:50 local time, inhabitants near the border between the States of Hidalgo and Puebla, located in central Mexico, witnessed a phenomenon similar to the one in Curuça: they heard a strong blast overhead, felt seismic tremors, and roof and windows vibration, but only a few persons saw a bolide. This was the first time Karina Cervantes de la Cruz, Eduardo Gómez, and Guadalupe Cordero went to field to investigate the fall of a cosmic object. After they interviewed more than 100 persons in 13 towns, they found only 12 persons that saw the bolide. With the information obtained from them, Karina Cervantes used a Wulff (equal-angle) stereo net to delimit the path of the meteor, finding that the object fall followed an azimuth between 55° and 90° with an angle to the horizontal between 16° and 47°. Comparing Curuça inhabitant narrations to Puebla-Hidalgo's, it is possible that the energy of the explosion was several kilotons. In addition, they realized that people fear these events and that civil protection and military agencies spend resources looking for something that can hurt people (Cordero et al. 2011).

One year later, on February 22, another object fell oriented more to the north-west, around the border between Zacatecas and Aguascalientes states. Again, people were afraid about this phenomenon, and it was more difficult to find witnesses of this event; that was the reason Cordero, Cervantes-de la Cruz, and Gómez started to plan the creation of the Mexican Meteor Network (or "Citlalin Tlamina" that means "meteor" in Nahuatl). With the support of CONACYT and the initial advice of Josep Trigo and José María Madiedo to choose appropriate cameras and lenses, the Citlalin Tlamina team started to work on the Network. The objectives of Citlalin Tlamina are to (a) study the fragmentation of meteoroids and small bodies in the Earth's atmosphere, (b) determine values of important parameters used to model the dynamics of objects through the atmosphere, (c) understand the interaction between the atmospheric shock wave and the surface, and what information this process provides about the height and energy of the explosion, (d) determine trajectories and orbital parameters of cosmic objects, (e) recover meteorites to study them, and (f) give information to population and Civil Protection Department to get its help and to avoid fear among inhabitants. Moreover, they want to install observational stations that can be functional, easy to manage, and to maintain and that operate without the help of an operator near them. One of the first thing was to design a platform to support video cameras (Whatec 902). The first design was very large and difficult to manage (Cordero-Tercero et al. 2016). With the collaboration of Fernando Velázquez Villegas, Alejandro Farah Simón, and Rodrigo García Fajardo, several designs were made. After several attempts, they got a compact, easy-to-manage platform made from 27 aluminum pieces that put together like a LEGO set. This platform is stable under vibrations (García Fajardo 2017).

During the same time of design and construction of the platform, the group also worked on (a) understanding the use of UFOCapture, the software used to record meteors; (b) understanding the way to find the path of a meteor from data of two or more stations; (c) programming data mentioned in (b) to determine pre-entry orbit, path through the atmosphere, and possible area to recover meteorite(s); (d) checking how to obtain meteor spectra; (e) testing how to recognize seismic record from air shock waves; (f) thinking how to protect the video cameras at the same time that keeping them at optimal conditions of temperature avoiding the point of dew. In addition, Citlalin Tlamina can be used to collect other kind of data such as clouds, bird migrations, and lightning. Much work needs to be done, but their creators expect to solve soon the technical problems they have. This project will permit us to study the sky with feet on surface at a low cost.

2.4 Mars: The History of the Search for Organics

Mars is a red, cold desert, but it was not always like that. In the past, 3900 billion years ago, it had liquid water on its surface, therefore a thicker atmosphere and a surface temperature above freezing (see Chap. 1). Comparing Earth and Mars at that time, they look very similar with active volcanoes, liquid water on the surface, and an atmosphere composed by carbon dioxide and nitrogen. On Earth, life emerged during that time; did the same happen on Mars? Part of Mars' exploration has focused on understanding its past and present geologic conditions and their habitability potential. Several scientific controversies have resulted from such research; in this section, we will focus on one that changed NASA's strategy to study organic matter on Mars and was starred by the Mexican scientist, Rafael Navarro-González.

In 1976, two probes landed on the Martian surface, *Viking 1* and *2*. They carried the first experiment to search for organic matter on Mars. Organics are carbon compounds that can be related to life, or not. Their presence in a planetary surface is interesting to astrobiology regardless of their source. When they do not come from life, they can be precursors of processes that result in the origin of life. For the Viking landers, analysis to detect organic matter was linked to the search for life (see Chap. 1). Although the experiments showed contradictory results that seem to indicate the presence of life, organics were not detected, and the conclusion was that there was no life at the Viking's landing sites. Mars was abandoned for 20 years and a new search for organics for more than 35 years; thus, there was enough time to think about the experiments and find out that maybe we missed something.

2.4.1 Mars on Earth: Clues on the Search for Organic Matter

The story starts in the Atacama Desert, the most arid place on Earth with temperatures between -2 °C and 32 °C, located at the south of Peru and north of Chile. Different from other deserts, its central region has no life because it never rains, and the air is totally dry. In that sense, it is similar to Mars, else the soil is oxidized, although not with iron oxides but sulfur and nitrogen oxides. This is what astrobiologists call an analog site, a place on Earth that shares some physical or chemical characteristics of other planetary body (see more in Chap. 1). Even when there is no life, organics arrive to the Atacama Desert carried by the wind. In 2001, Rafael Navarro-González and Christopher McKay (NASA Ames) took their instruments to Atacama and analyzed the soil to detect organics. Navarro-González used the same technique of the *Viking* landers. Organics were no detected although they knew they were there. Other researchers joined the quest and the same technique was repeated in other analog sites: the Antarctic Desert; the Lybian Desert; the Mojave Desert; Rio Tinto, Spain, and Panoche Valley, California. For most of them, the *Viking* technique did not work for detecting organics. For the *Viking* analysis, samples were heated at 500 °C, Navarro-González and collaborators used the same temperature and ran another set of experiments at 750 °C. Higher temperature allowed to detect organics in several sites, indicating that an oxidizing reaction during the analysis was responsible for not detecting organics on Mars (Navarro-González et al. 2006).

After the publication of these results, the head of the team that build the instrument that analyzed the Martian soil was not happy. The Viking's analysis instrument was a huge achievement when it was built because the technology existed, but it had to be small enough to fit in the *Viking* landers. All the miniaturization work and the testing of the instrument were led by Klaus Biemann. His answer to the results of Navarro-Gonzalez and collaborators was published in 2007 (Biemann 2007) in the same scientific journal. He claimed that "the experiments reported by Navarro-Gonzalez et al. (2010) are poorly conceived, executed, and interpreted...." The team did not respond, but that was not the end of the story. In 2009, it was announced that *Phoenix*, a lander that arrived at the Martian Arctic, found perchlorates on the soil. Navarro-González' team reproduced their 2006 experiments but now including these chlorinated compounds in the samples (Navarro-González et al. 2010). They found that after heating at 500 °C, organic compounds reacted with perchlorates producing chloromethane, a molecule that was detected in the *Viking* analysis. The only problem is that chloromethane was used to clean the *Viking* instruments, so there were two possibilities: *Viking* detected martian organics or detected its own cleaning product. Biemann and Bada (2011) argue that cleaning products and not organic compounds were detected by *Viking* instruments, Navarro-González and McKay (2011) replied to every one of the critics and stand by their results.

2.4.2 Curiosity Is What Is Needed

In 2006, NASA was planning *Curiosity*, a rover the size of a small car, with several instruments to study Mars, including search for organics. Navarro-González and collaborators results changed the planned instruments for this purpose, and Rafael was included in the *curiosity* team. The rover landed on Mars in 2012 at the Gale Crater formed when a meteor hit Mars, about 3.5–3.8 billion years ago, and there was evidence of liquid water in the past (Mars Exploration Program and Jet Propulsion Laboratory). In 2014, the first results that hinted to the presence of organic compounds in Mars, as predicted by Navarro-González and collaborators in 2010, were published. The detected organics contained chlorine acquired by reactions during the analysis process (Ming et al. 2014; Freissinet et al. 2015). *Curiosity* continued his work on Mars, and in 2018, a team lead by Jennifer L. Eigenbrode (Eigenbrode et al. 2018) announced the detection of sulfur-bearing organic matter preserved in a 3-billion-year-old mudstone. Presence of organics in Mars is now confirmed beyond doubt, where it comes from is something else. Several sources are possible, from meteoritical materials to geologic processes. In any case, we can say that all elements for the origins of life were in Mars, about four billion years ago.

2.5 Icy Worlds: Europa and Enceladus

An exciting debate connected to Astrobiology was published in *Nature* journal: "Mars habitability" as the focus of interest (Lowell 1908). This was due to Mars atmosphere ability to retain water. The atmospheric centered idea of habitability was preponderant; indeed, it brought up the "habitable zone" concept, which is defined as the circumstellar region where a terrestrial planet (iron core and silicate mantle and crust) with an atmosphere can retain liquid water on its surface (Segura 2018).

During the late 1970s, with the first observations of hydrothermal systems at the Galapagos Spreading Center, the "habitable zone" concept was expanded (Dick 2013). The concomitant *Voyager* spacecraft discovery of a very young state of the icy satellite's surfaces in 1979, with very few impact craters on Europa surface, promoted the development of theoretical studies (e.g., Squyres et al. 1983), which supported the existence of a liquid water ocean, under an ice crust, that is still liquid nowadays (Scipioni et al. 2017).

2.5.1 Europa

The Jupiter Galilean system (Jupiter and the satellites Io, Europa, Ganymede, and Callisto) has been subject to intense study since its discovery by Galileo Galilei in 1601. Galilean satellites are far from being inactive because of their viscoelastic properties combined with the fact that they undergo periodic tidal deformations. However, it has been calculated that radiogenic heating, a product of radioisotope decaying, was enough to differentiate Europa leading to a water layer (about 100 km). Tidal heating, resulting after satellite deformation, is responsible for keeping most of its thick water crust melted, which maintains its global ocean (Nimmo and Pappalardo 2016). The dissipation of deformation energy is a valuable source of heat in satellites from outer Solar System and help to explain, for example, Io's volcanic activity and the plausible hydrothermal systems on Europa and Enceladus. Therefore, the warming of icy satellites as a sequel of tidal heating is valuable enough to redefine the habitable zone concept.

The geological features obtained by *Galileo* orbiter flybys, in 1997, informed immediately about the occurrence of tectonic, cryovolcanic, and diapiric processes (Cordero and Mendoza 2004) in Europa. The discovery was enough to support a two-year extension of the *Galileo Europa Mission* (GEM) to 1999 in the search of a consistence with the presence of liquid water beneath the water ice crust. At first, the satellite's magnetic field induced by Jupiter's was the conclusive evidence for an Europan ocean because it is consistent with a global electrically conducting shell and suitable to be salt water (Nimmo and Pappalardo 2016). *Hubble Space Telescope* (HST) observations of Europa surface, performed in 2012, showed the first evidence of water plumes of a tidal origin, because of their magnetic and plasma wave signatures (Table 2.2). Afterward, the HST water plumes interpretation was confirmed with data available from the *Galileo* Magnetometer and the *Stratospheric Observatory for Infrared Astronomy* (SOFIA) Spectrograph (e.g., Roth et al. 2014). Ocean physical–chemical features such as pH, salt composition and concentration, and redox state are also a matter of debate. At first, *Galileo* Near-Infrared Mapping Spectrometer (NIMS) pictured a view of a hypersaline ocean because of the finding

Table 2.2 Europa and Enceladus habitability-linked features that promote them as the best candidates, among icy satellites, for habitability. Despite that some properties constraint habitability requirements (liquid water, chemical energy source, and organic carbon molecules), Europa and Enceladus are still in the astrobiology roadmap

	Europa	Enceladus	References
Radius and layers (km)	1565.0 ± 8.0 Water cortex (106), mantle (980), and core (475)	252.1 ± 0.2 Water cortex (62) and core (190)	Vance et al. (2018)
Foundation for a liquid water ocean	The inference of eruptive water emissions with the Hubble Space Telescope Imaging Spectrograph (STIS), though they are infrequent when compared with Enceladus jets. Galileo spacecraft detection of a Jupiter-induced magnetic field and clearly distinguished from an internal field	The discovery of geysers and thermal emission at Enceladus' south polar region (along the tiger stripes) linked to tidal forces. The detection of silica-rich particles originating from the plume suggests a hydrothermal rock-water interaction beneath the ice layer	Roth et al. (2014) Nimmo and Pappalardo (2016) Postberg et al. (2011)
Ocean salts	$MgSO_4 >> Na_2SO_4$, Na_2CO_3 or $KCl > NaCl$ Sulfates >> carbonates > chlorides	$NaCl > NaHCO_3 + Na_2CO_3$ Chlorides > carbonates >> sulfates	Hibbitts et al. (2019) Glein et al. (2015)
Ocean salinity (g kg^{-1})	282	3–8	Hand and Chyba (2007) Zolotov (2007)
Ocean pH	It is not expected an extreme pH Mg amount dependent	Between 8 and 12	Johnson et al. (2019) Glein et al. (2015) Zolotov (2007)
Are organics confirmed?	No, however, organics have been used as a possible explanation to Galileo Solid-State Imager (SSI) data	Yes, hydrocarbons (C_2 to C_5 and C_7 to C_{15}) and ammonia have been detected in the plume	Hand and Carlson (2015) Postberg et al. (2018)
Laboratory chemical simulations	Precipitation of inorganic iron-sulfide membranes under hydrothermal solutions and generation of a membrane potential	Production of gas species (H_2, NH_3, CO_2, CH_4, N_2) by serpentinizing-like reactions	Sekine et al. (2015) Barge and White (2017)
Example of suitable life style	Simple (unicellular), hydrogen-sulfate reduction	Simple (unicellular), hydrogen-methanogenesis	Zolotov and Shock (2003) Taubner et al. (2018)

(continued)

Table 2.2 (continued)

	Europa	Enceladus	References
Laboratory biological simulations	Halotolerant bacteria growth under varying sulfate and salt concentrations	Methanogenic archaea growth under hydrothermal-like conditions (varying fluid composition)	Avendaño et al. (2015) Taubner et al. (2018)
Environmental analogs	Vostok Lake, East Antarctic ice sheet Tirez Lake, Spain Borup Fiord Pass, Canada Lost City hydrothermal field, Mid-Atlantic Ridge Samail ophiolite, Sultanate of Oman	Wadi al Natrun, Egypt Kulunda Lake, Russia Mono Lake, USA Rincon de Parangueo, Mexico Lost City hydrothermal field, Mid-Atlantic Ridge Samail ophiolite, Sultanate of Oman	Lorenz et al. (2011) Glein et al. (2015)
Habitability status	To be established by NASA's Europa Clipper mission	Sufficiently established by Cassini-Huygens mission	Hendrix et al. (2018)
Next confirmed mission	NASA's Europa Clipper ESA's JUICE (Jupiter Icy moons Explorer)	None	

of hydrated minerals, while magnesium sulfates and sodium carbonates provided the closest match to NIMS reflectance spectra. Recently, it has been stated that chloride salts were omitted because of *Galileo* NIMS sensitivity bias, to clear statements new spectra were obtained with HST and sodium chloride (NaCl) was confirmed; consequently, the actual statement is that salt water system (SO_4-Cl content) relies on the hydrothermal circulation extent (Trumbo et al. 2019).

In 2002, Lilia Montoya performed the first experimental test for the suitability for life in Europa at the Instituto de Ciencias Nucleares (ICN), Universidad Nacional Autónoma de México (UNAM). The research group was headed by the chemist and astrobiologist, Rafael Navarro-González; he and Montoya worked in collaboration with the astrophysicists and astrobiologists Wanda L. Davis and Christopher P. McKay (NASA Space Science and Astrobiology Division). The experimental approach evoked hypothetical hydrothermal systems with a pressurized fluid-circulating reactor. Covered temperatures included those common of low temperature (100 °C) and high temperature hydrothermal systems (375 °C), other experimental variables included mineral and fluid composition. Retrieved molecules from the gas phase were identified as hydrocarbons, a diverse set of C_2 to C_7 isomers (Montoya et al. 2002). This approach underlined the likely of hydrothermal systems to be a source of chemicals and energy to organisms (by hydrocarbon cleavage, hydrogen oxidation, and methane synthesis) in Europa seafloor.

Later, as part of Montoya's postgraduate investigation, she studied the life associated with the Europa analog system Tirez Lake, an athalassohaline lake located in Castilla La Mancha, Spain (Fig. 2.4). The experimental research was concreted thanks to the mentoring of the geomicrobiologists and astrobiologists, Ricardo Amils, and Irma Marin and the association of laboratories from the following research institutions: Centro de Biología Molecular Severo Ochoa (CBM, CSIC),

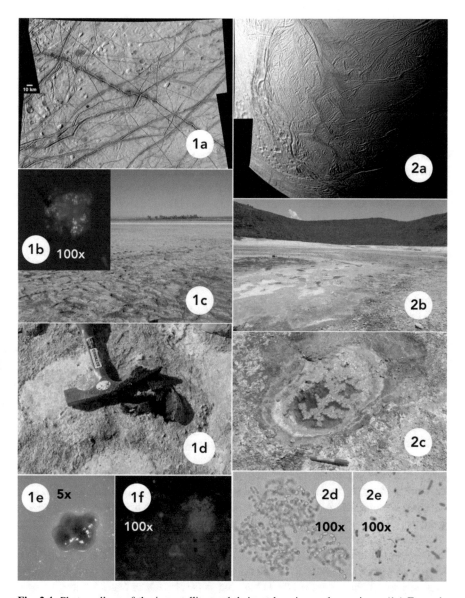

Fig. 2.4 Photo collage of the icy satellites and their analog sites and organisms. (1a) Europa's Rhadamanthys Linea captured by Galileo spacecraft. The region is covered by characteristic features, such as ridges and lenticulae. (2a) Saturn's moon Enceladus Great Southern Land acquired by NASA's Cassini spacecraft, the "tiger stripe" fractures have been green colored. (1c, 1d) Europa analog environment Tirez Lake in Castilla La Mancha, Spain. (2b, 2c) Enceladus analog environment Rincón de Parangueo in Guanajuato, Mexico. (1b) Epifluorescence micrograph of a Tirez sediment sample treated with a modified version of the fluorescent in situ hybridization protocol (CARD-FISH). Hybridization treatment highlights the Bacteria cells in green. (1e) Amplified view of Tirez sulfate-reducing colonies. When the colonies exhibit a dark precipitate (iron sulfide, FeS$_2$), the sulfate-reducing activity is confirmed. (1f) Epifluorescence micrograph of a Tirez sediment sample treated with FISH protocol to highlight the sulfate-reducing bacteria in orange. (2d) Bright field micrographs of the Cyanobacteria *Aphanotece* sp. and (2e) an anaerobic isolate. Both organisms were found in Rincon de Parangueo water samples. (1a) Image credit: NASA/JPL-Caltech/ University of Arizona. (2a) Image credit: NASA/JPL/Space Science Institute. (1b-1f) Image credit: L. Montoya. (2b-2e) Image credit: M.E. Hernández-Zavala

the Unidad de Microbiología Aplicada at the Facultad de Ciencias (FC, Universidad Autónoma de Madrid), and the Centro de Astrobiología (CAB, INTA-CSIC). Tirez microbial inventory supported the idea of viable life based on methanogenesis and sulfate reduction in Europa ocean because this lake has already been validated as a reliable analog for Europa due to a low Cl/SO_4 rate and the presence of magnesium and sodium as main cations (Lorenz et al. 2011). A subsequent research took into account the ephemeral condition of Tirez Lake to analyze the resilience (ability to recovery after a disturbance) of methanogenic and sulfate-reducing populations, and was developed in a collaboration with the evolutionary biologist, Irma Lozada-Chávez, from the Interdisciplinary Center for Bioinformatics at the Leipzig University, Germany. They showed that Tirez Lake harbors resilient halophilic (extremophiles characterized by their preference for hypersaline conditions) sulfate reducers but not all methanogens share this condition. Also, bioenergetics play an explanation to the hydrogen-methanogenic un-resilience since a low energy yield. Therefore, salinity is a significant parameter to be considered in evaluating the habitability in Europa (Montoya et al. 2011).

After the "burst" of information provided by *Galileo* Mission, it became pertinent to define biomarkers ad hoc to Europa; so compatible solutes were an obvious option. The obviousness of compatible solutes, i.e., low molecular weight and organic molecules responsible for cell internal stability, came from the fact that they are widespread in those extremophiles linked to icy satellites studies (halophiles, psychrophiles, and hyperthermophiles), including methanogens and sulfate reducers, and these are present in all domains of life. As a consequence, in collaboration with the geophysicist, Javier Ruiz, from the Facultad de Ciencias Geológicas at the Universidad Complutense at Madrid, they proposed the search for compatible solutes in the specific Europan features "chaos" and "lenticulae" regions because their origin has been explained as a result of a diapiric process, i.e., plumes of warm ice ascending through the ice crust and disrupting the surface (Ruiz et al. 2007). In Mexico, in collaboration with the planetary chemist, Sandra I. Ramírez-Jiménez, from the Centro de Investigaciones Químicas, Universidad Autónoma del Estado de Morelos (CIQ, UAEM), they started the strategy of analyzing the compatible solutes from cultured bacteria. They observed that prokaryotes synthesize compatible solutes and grow (tolerate) under the common salt sodium chloride (NaCl) and surprisingly when changing the NaCl to sulfate salts (Na_2SO_2 or $MgSO_4$), displaying a particular correlation between the chemical identity of salts and that of the accumulated compatible solutes (Avendaño et al. 2015).

2.5.2   Enceladus

Cassini mission performed a close flyby above the south pole of the Saturn's small moon Enceladus in 2015. Thereafter, by displaying jets of water vapor emanating probably from the southern large fractures named as "tiger stripes," this icy and active moon gained a place in the astrobiology roadmap. *Cassini*'s Magnetometer corroborated the ocean profile. Consequently, the subject of interest was transferred from a tidally induced plume to fields, such as parallel evidences of a liquid water ocean, ocean composition, and suggestions of hydrothermal activity. Two *Cassini* detections were substantial to infer the hydrothermal activity in the interphase cortex-mantle: silica nanoparticles and molecular hydrogen at disequilibrium abundance (Vance 2018). Enceladus' ocean has already been described as global, instead of a huge aquifer, and its salt content is certain when compared with Europa's. Enceladus' observations have inferred an alkaline ocean and an anion content far from sulfate (SO_4) but close to chloride-carbonate (Cl-CO_3) (Glein et al. 2015).

Given the close understanding of Enceladus "oceanographic characteristics," the closely resembling environments were immediately the next step: search the overlapping hypotheses of habitability, biosignature, and biomarker detection. Probably, the best terrestrial analog environment alike to Enceladus is the Lost City at the Mid-Atlantic Ridge. Some hydrothermal systems have been proposed, other than Lost City, as long as they fit with the low temperature (i.e., a temperature $\leq 100\ °C$) and neutral to alkaline pH (Ramírez Cabañas 2017). Low-temperature mid-ridge hydrothermal systems are also known as white smokers; they are very distinctive for being serpentinization sites. During serpentinization reactions, ultramafic minerals react with water and release molecular hydrogen, leading to hydrothermal fluids with reducing conditions. Molecular hydrogen (H_2) is a powerful reducing agent and source of energy for chemolithotrophs when an oxidant such as carbon dioxide, oxygen, sulfate, or iron (III) is present. Therefore, it has been growing the consensus that Enceladus might be habitable for chemotrophic organisms, i.e., capable of extracting energy from chemical reactions instead of light as plants or cyanobacteria. Besides making hydrogen, serpentinizing reactions act as scaffold to complex organic synthesis, examples are Fischer-Tropsch type reactions (FTT). The FTT reactions have been used to explain the finding of organic macromolecules emitted from Enceladus plume ejects (Postberg et al. 2018).

Continental soda lakes, such as Wadi al Natrun (Egypt) and Mono Lake (USA), have been proposed as analog and accessible sites to inquire Enceladus' habitability. The Mexican maar crater lake Rincon de Parangueo in Guanajuato was proposed as an analog system to Enceladus after a collaboration with the team of researchers from the Instituto Potosino de Investigación Científica y Tecnológica (Ipicyt) in Mexico: the geophysicist Alfredo Ramos and the geoscientist M. Elizabeth Hernández-Zavala from the División de Geociencias Aplicadas and the biotechnologist Ángel G. Alpuche-Solís from the División de Biología Molecular (Montoya et al. 2017). This analogy contemplated especially the research developed by the geoscientist, Hernández-Zavala (2016), regarding the hydrogeochemical and microbial characterization of this maar crater lake. M. Elizabeth Hernandez-Zavala reported the coexistence of low redox values in the water in concomitance with diverse archaea and bacteria (Fig. 2.4). These data offered astrobiological implications because from the bioenergetic point of view, the anaerobic organisms are the most constrained extremophiles; therefore, their finding in Rincon de Parangueo widens the habitability window in the Saturn satellite. Theoretical and experimental strategies to ascertain Enceladus habitability have been already explored (Table 2.2).

However, to accomplish the purpose of being objective in assessing such habitability and in accordance with the concept of quantitative habitability plentifully developed by the biogeochemist and astrobiologist Tori M. Hoehler (NASA Ames Research Center), Lilia Montoya proposed this quantitative approach to Enceladus habitability. This inquire has been developing at the Facultad de Ciencias Biológicas (FCB) at the Universidad Autónoma del Estado de Morelos (UAEM) with the involvement of the biologist Carlos A. Soriano and the marine biologist Elva Escobar-Briones from the Instituto de Ciencias del Mar y Limnología (ICMyL, UNAM). Fortunately, data let them to approach a habitability index, i.e., a quantitative value that allows the comparison between analog sites, e.g., Europa vs Enceladus (Soriano-López et al. 2017). In advance, their estimates, constrained by salinity, pH, and hydrothermal vent temperature and pH, have revealed that Enceladus has a higher habitability index compared to Europa. Among the icy satellites, Enceladus promises a better understanding of the distribution of life in Universe, indeed a search-for-life mission to it will be a priority in the Solar System exploration. Undoubtedly, efforts devoted to gain a better understanding of both satellites will be well rewarded.

2.6 Titan

Titan is the biggest of the 62 confirmed satellites of Saturn, second in size among the satellites in the Solar System, and the only one that holds a dense, cold, and dynamic atmosphere in which a continuous abiotic synthesis of organic compounds occurs (Fig. 2.5a). The major atmospheric constituent is molecular nitrogen (N_2), the same compound that makes the majority of the terrestrial atmosphere, followed by methane (CH_4), the smallest of the organic molecules. Titan is the only other world aside from Earth with a permanent presence of liquids on the surface, and an active hydrological cycle running, but with methane and ethane taking the place of water (Coustenis 2014) because of its unique conditions of average surface temperature of about 94 K (-179 °C) and a surface pressure of 146.7 kPa (1.45 atm) (see Chap. 1 for more on Titan potential habitability).

Titan is a satellite of astrobiological importance as many of the processes that occur on a planetary scale can be considered close analogs to ancient terrestrial processes that can be monitored remotely and hopefully in situ.

2.6.1 The Voyager and Ground-Based Observations Era

Exogenous and endogenous energy sources promote the chemical transformation of the dense and mild-reducing atmosphere of Titan originating gaseous hydrocarbons and nitriles, which after different polymerization processes yield solid aerosol particles that grow by a variety of mechanisms and fall to the surface.

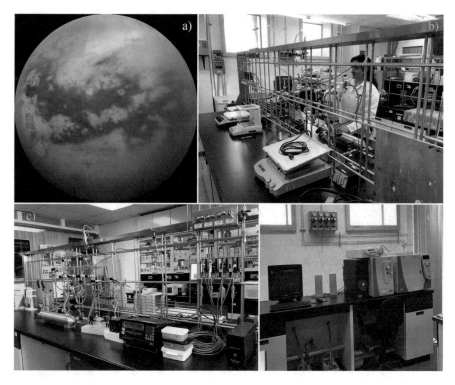

Fig. 2.5 (**a**) Titan in false colors. Image credit: NASA/JPL. (**b** and **c**) Views from the infrastructure at Laboratorio de Simulación de Ambientes Planetarios (LSAP) at CIQ, UAEM. (**d**) Analytical platform at LSAP. (Images (**b–d**) credit: Sandra I. Ramírez Jiménez)

The study of the organic chemistry happening on Titan requires direct observations, generation of models, in situ sampling, as well as laboratory simulations and analysis of the synthesized products, all of these from atmospheric and surface processes. As part of a strategy to run experiments in conditions as close as possible to Titan's conditions, a laboratory devoted to the simulation of Titan's atmosphere was created at Centro de Investigaciones Químicas (CIQ) at the Universidad Autónoma del Estado de Morelos (UAEM) (Fig. 2.5b–d). Prior to the arrival of Cassini-Huygens, several experiments were performed in collaboration with laboratories at the Universidad National Autónoma de México (UNAM) and Université Paris-Est Créteil (UPEC).

One of the first systematic studies of Titan's atmosphere was performed at Laboratoire Interuniversitaire des Systèmes Atmosphériques (LISA) at UPEC, where an experimental setup was developed to study the chemistry of Titan's upper atmosphere at low pressure under a liquid nitrogen temperature regime. Calculation of energy yields using a simple mixing atmospheric model was performed (Coll et al. 1999). A total of 44 hydrocarbons and 26 nitrogen-containing gas-phase compounds were synthesized. The detection of butynedinitrile (C_4N_2) was reported for the first time. A yellow solid, analog to Titan's hazes, was also recovered. Its spectroscopic characterization indicated absorption bands in the infrared corresponding to C ≡ N bonds; pyrolysis analysis reported some benzene derivatives, and a few hydrocarbons and nitriles, while electron microscopy observations showed that the haze had a quasi-homogeneous spherical distribution (Coll et al. 1999).

In the meantime, the role played by corona discharges developing among predicted methane clouds in Titan's troposphere was studied at Instituto de Ciencias Nucleares (ICN) at UNAM. Detailed argumentation was presented to consider the likelihood of electrical activity in Titan. The experiments were done at room temperature and high pressure. The results demonstrated that even weak electric fields could form a variety of saturated hydrocarbons and nitriles, some of them underestimated by photolysis or radiolysis experiments, together with molecules of astrobiological importance such as hydrogen cyanide (HCN) (Navarro-González and Ramírez 1997). The first evidence of methane clouds in Titan was presented in 1998 (Griffith et al. 1998) and was confirmed by different techniques in the following years (Griffith et al. 2000; Roe et al. 2002; Brown et al. 2002; Gibbard et al. 2004).

To have a better representation of Titan's scenario, a quantitative comparison of the products arising when a simulated atmosphere was irradiated by different energy sources was performed. A laser-induced plasma (LIP) represented the entrance of high-velocity meteors into Titan's atmosphere. Gamma radiation simulated the effect of Saturnian electrons going through Titan's stratosphere and of galactic cosmic rays ionizing neutral molecules in their way to the surface. Arc and corona discharges were used to simulate the development of electrical activity within the tropospheric methane clouds (Ramírez et al. 2001). The evaluated energy sources demonstrated to be good abiotic nitrogen fixers as nitriles were synthesized in all cases. Corona discharges preferentially produced saturated and methyl-branched hydrocarbons, while conjugated hydrocarbons and aromatic compounds were found in arc discharge experiments. The LIP originated mainly alkynes, unsaturated nitriles, and benzene derivatives. Gamma rays produced saturated and unsaturated hydrocarbons, as well as aromatic compounds.

An important outcome from these experiments was a light-yellow solid deposit observed in the reactor walls for all of the energy sources, except for the gamma radiation. This solid can be considered an analog of the aerosols that form the haze layers detected in Titan's stratosphere, also known as *tholins* in the jargon of experimentalists. The characterization of the aerosol analogs by infrared spectroscopy confirm the C ≡ N absorption observed in the analogs synthesized at LISA, but show additional signatures attributed to aliphatic -CH$_3$, -CH$_2$-, C=C, C ≡ C bonds; C-N amines and imines bonds; and aromatic C-H, =C-H bonds. All these types of bonds were confirmed by nuclear magnetic resonance spectroscopy (Ramírez et al. 2004).

2.6.2 The Cassini-Huygens Era

The *Cassini-Huygens* space mission involved a collaboration between the American, European, and Italian Space Agencies to send the *Cassini* probe (CP) to study Saturn, its rings and natural satellites, as well as the *Huygens* lander (HL), the first spacecraft ever to successfully land on a moon in the outer Solar System, and the farthest from Earth. Their scientific instruments revolutionized our understanding of the Saturn system and helped in lifting Titan's veil providing a wealth of new information.

The Cassini Plasma Spectrometer (CPS) on board the HL detected heavy negative ions (10,000 Da), meaning high molecular weights for the organics contributing to the haze layers found in the stratosphere (Waite et al. 2007; Wahlund et al. 2009; Coates et al. 2009). The results from the gas chromatograph and mass spectrometer (GC-MS), and the aerosol collector and pyrolizer (ACP) experiments on HL suggested that Titan's haze was made of refractory organics, containing C, N, and H atoms and releasing, when pyrolyzed at 600 °C, HCN and NH_3 (Israel et al. 2005). These results strongly suggest that the complex organic materials constituting the aerosol particles are similar to some of the Titan's aerosol analogs produced in terrestrial laboratories, thus validating their role as analogs.

The existence of a subsurface liquid water ocean containing a few percent of ammonia was one of the most striking results from the Cassini-Huygens mission. Several lines of evidence suggest that Titan's interior contains ammonia (NH_3) mixed with water, in the form of a liquid layer below a rigid water ice crust that leads to the possibility that outgassing or erupting fluids could bring NH_3 to the surface (Tomasko and West 2010; Fortes 2000; Tobie et al. 2005; Lorenz et al. 2008; Béghin et al. 2009; Sotin et al. 2010). If this is the case, a chemical transformation of the aerosols on the postulated NH_3-H_2O mixtures introduced onto the surface as a cryovolcanic eruption, or as a result of crustal melting following an impact cratering event, is expected. To understand the role that aqueous ammonia plays in the chemical transformation of atmospheric aerosols laboratory analogs, a N_2:CH_4 (98:2) mixture was irradiated in a low-temperature continuous flow regime by a DC cold plasma discharge at LISA. The aerosol analogs were exposed to basic hydrolysis conditions (0.8–6.5 M NH_4OH) at 277, 253, and 96 K for 10 weeks. The hydrolysis products were identified by the CG-MS platform at CIQ after their transformation into volatile MTBSTFA derivatives. Seven amino acids and urea were detected demonstrating an interesting fate for the aerosols gravitationally deposited on Titan's surface, where in spite of the low temperatures, chemistry can lead to the abiotic synthesis of organic and biologically attractive compounds. What is more, based on an aerosol mass flux of the order of 2–3×10^{-14} g cm^{-2} s^{-1} (Tomasko and West 2010), one milligram of aerosols can be deposited over an area of 1 cm^2 on Titan's surface over 1000 years. If only 0.1% of these aerosols is transformed into amino acids, the outcome would be on the order of one microgram per cm^2, a quantity detectable in situ with current analytical technology (Ramírez et al. 2010).

2.6.3   The Future for Titan Exploration

It is true that when trying to find the answer to a specific question, more questions are generated, and Titan's environment is not the exception. To fully understand the global chemistry of Titan, laboratory protocols that support informed models to help to develop more effective in situ analytical instrumentation are required. New proposals in each front are now being prepared to document the likelihood of signatures of water-based or hydrocarbon-based life, as well as the habitability of Titan, one of the astrobiological jewels of the Solar System.

2.7   Planets Around Other Stars

Exoplanets are planets that orbit stars that are not our Sun. The word "exoplanet" is derived from the words *extrasolar* that comes from the Latin words *extra* (beyond) and *solar* (the Sun), and the Greek word *planet* (wanderer), which although has existed for many centuries, its meaning is still evolving (e.g., IAU resolution regarding Pluto on 24 August 2006). The names of exoplanets are given by the name of the star followed by a lower-case letter (starting with "b" for the first planet to be discovered around a given star, "c" for the second, etc.). For example, the closest star to our Sun is Proxima and the exoplanet around it is Proxima b (Anglada-Escudé et al. 2016).

The first known exoplanet system (PSR + 1257 12) was discovered in 1992 with three Earth-mass planets orbiting around a rapidly rotating neutron star or pulsar (remnant of a supernova explosion; Wolszczan and Frail 1992; Wolszczan 1994). It was discovered by measuring pulses (hence pulsar) that are beamed periodically every few milliseconds toward Earth. The deviation from a very precise period of these pulses is what allows us to know that these planets exist and from which we can measure their masses. Until then everything we knew of planets was based on our Solar System, and its discovery was tantalizing. However, at this point, we did not believe this discovery was relevant to the Solar System, since the planets revolved around a neutron star and not a "normal" star like our Sun. Furthermore, the pulsar planets would need to survive the supernova explosion or would need to form a second generation of planets, created from the material that is left over after the supernova.

It was a few years later that Mayor and Queloz (1995) discovered the first planet around a star like the Sun (51 Pegasi), by measuring the star's velocity changes in time due to the planet tugging on the star as the planet orbits around it. Modeling the change in the star's velocity allows the measurement of planet mass. This discovery blew our minds, because 51 Pegasi b is a gas giant exoplanet that goes once around its star in a little over 4 days. Mayor and Queloz earned half of the 2019 Nobel Prize in Physics for finding 51 Pegasi b, the first planet around a star like our Sun. We do not have examples of this kind of planets in our Solar System. This new and unexpected type of planet is named hot Jupiter as planets in this class have a mass like Jupiter, but orbit once around their star every few days receiving from it up to 10,000 times the irradiation that the Earth receives from the Sun (e.g., Demory and Seager 2011).

In the last decade, the number of known planets outside the Solar System has sky-rocketed. We went from celebrating the discovery of the first few hundreds of known exoplanets in 2010 (like the 98th exoplanet with a measured size, WASP-38 b; Barros et al. 2011) to having confirmed more than 4000 exoplanets (accessed on November 4, 2019; Schneider et al. 2011). The sizes of the planets are measured from the amount of starlight that the planet blocks as it passes in front of the star eclipsing it; such a planet is called a transiting planet, and when the planet crosses the star, it is called a transit. This boom has been the result of a worldwide effort encompassing the blossoming of facilities and research groups (e.g., Bakos et al. 2002; Pepe et al. 2004; Pollacco et al. 2006), the development of new strategies and analysis techniques (e.g., Siverd et al. 2012; Gillon et al. 2016, 2017; Talens et al. 2017), and the start of operations of new observing instruments and facilities (e.g., the first satellite dedicated to exoplanets NASA's Kepler was launched in 2009, Borucki et al. 2010; Quirrenbach et al. 2014; Ricker et al. 2014). In fact, in 2011, a group of Mexican astronomers (Curiel et al. 2011) applied a genetic algorithm to velocity data of the star Upsilon Andromedae and were able to newly identify a fourth planet in the planetary system. Upsilon Andromeda d is very similar to our own Jupiter, with a mass of at least 1 Jupiter mass and taking ~10.5 years to orbit once around its star (Jupiter takes 11.86 years to go around the Sun once). Mexican astronomers have continuously participated in large consortiums in the finding of new exoplanetary systems and studying them (e.g., Gómez Maqueo Chew et al. 2013a, b; Barragán et al. 2018).

Thus, understanding the physical processes that are involved in the formation and evolution of our own Solar System was not enough to explain the diversity of exoplanets and environments in which they have been found. A clear example, besides hot Jupiters, is the population of planets with masses around 10 Earth masses that have between 1 and 4 times the diameter of the Earth. Depending on whether they are expected to have rocky compositions, they are called Super-Earths, or if they are expected to have significant atmospheres of hydrogen and helium, they are called Mini-Neptunes (Howard et al. 2010; Fulton and Petigura 2018). This population is not known to exist in our Solar System, but among the known exo-planets, Super-Earths and Mini-Neptunes are the most numerous (Petigura et al. 2013). One of the most important exoplanet discoveries was the discovery of Kepler-16, the first planetary system around two stars (Doyle et al. 2011); finally science caught up to science fiction's Tatooine from Star Wars.

So far, the field of exoplanets has been without a doubt observationally driven, meaning that breakthroughs in scientific knowledge have been pioneered by dedi-cated astronomical observations. To this end, the first telescope in Mexico fully dedicated to exoplanets was installed at the Observatorio Astronómico Nacional in San Pedro Mártir, Baja California, in late 2018 (see Fig. 2.6). This Swiss-Mexican project, named SAINT-EX (Search And characterIsatioN of Transiting

Fig. 2.6 On the left, the SAINT-EX dome is shown at sunset just before starting astronomical observations, and the telescope can be seen through the shutter. On the right, the SAINT-EX telescope (in yellow) and the equatorial mount (in light blue) are shown from inside the dome. Both photos were taken over the summer of 2019 at the Observatorio Astronómico Nacional de San Pedro Mártir, in Baja California, México by E. Cadena

EXoplanets), consists of the installation and operation of a fully robotic and remotely operated telescope with a 1-meter primary mirror on an equatorial mount equipped with a state-of-the-art CCD camera (González et al. 2018). The SAINT-EX has been in the testing phase throughout 2019, and it has begun astronomical observations to search for new transiting exoplanets around the smallest, least massive stars in our Solar Neighborhood. The SAINT-EX team is led by B.O. Demory, as Principal Investigator, and by Y. Gómez Maqueo Chew, as Project Coordinator, and is composed of D. Queloz (recipient of the 2019 Physics Nobel Prize), W. Benz, F. Bouchy, K. Heng, M. Gillon, and L. Sabin, as well as students and postdocs. The team will search for new exoplanets by measuring the light from the target stars as a function of time to identify periodic dimmings due to the transiting exoplanets crossing in front of their stars. Any new exoplanet discoveries from SAINT-EX will provide optimal targets to observe with the next generation of space telescopes (e.g., NASA's *James Webb Space Telescope*) and explore the atmospheres of rocky planets outside of our own Solar System.

2.8 Habitable Worlds and Signatures of Life

In 2003, Antígona Segura was working as a postdoctoral researcher in Penn State University in the group lead by James Kasting, an expert on planetary habitability. Her first work was to make the experiment of placing Earth around other stars to understand how the signals of life on Earth would change. They were not really moving Earth but using models to simulate a planet just like ours irradiated by two stars, one hotter than the Sun, an F star, and other one cooler, a K star. Of course, they used the Sun (a G star) to know if their simulations were reproducing our planet's atmosphere. Different stars mean different amounts of ultraviolet (UV) light for those "earths," and thus changes on the destruction and creation of chemical species in the atmosphere. The chemical species they were interested in were the ones produced by life because they may be useful in the future to detect life on other planets, and those chemicals are named biosignatures (see Chap. 1).

Earth's atmosphere contains enough oxygen produced by photosynthetic bacteria to be detectable by telescopes, but oxygen is not always a biosignature (see Chap. 1), for example, this compound is present in small concentrations in the atmospheres of Mars and Venus produced by the breakup of CO_2 molecules by UV without the intervention of life. Back to the experiment Segura was conducting as her first job in Penn State University, they knew hotter stars produce more UV than the cooler stars, then all the compounds produced by splitting molecules with this radiation would change their concentrations in the atmospheres of the earths they were simulating. Results published in 2003 (Segura et al. 2003) showed which compounds of the simulated earths would be detectable in planets around those stars if chemicals were produced by life at the same rates than on Earth. To extend this work, Segura chose the coolest stars, M dwarfs. These stars were not studied before mainly for two reasons, one is that the model crashed when Segura used the flux of these stars and the other is that for most M dwarfs, UV light was not detectable. There were measurements only for those identified as "active". This means that one part of their stellar atmosphere, the chromosphere, produces more X-rays and UV than other stars. This kind of radiation may make a planet around an M dwarf an unsuitable place for life, so non-active stars would be better, but there were no measurements available of UV emission for those stars classified as non-active. What they did was to take a hypothetical UV emission calculated by a numerical model to represent a non-active M dwarf and the emission of a very active one, called AD Leonis. And then the surprises started.

In 1965, Lovelock suggested that the simultaneous presence of O_2 and methane (CH_4) in a planetary atmosphere may be a good indicator for life. The reason is that a series of chemical reactions starting with O_2 would destroy CH_4. In our planet, life produces one billion kilograms of CH_4 every year, but its abundance in the atmosphere is only 1.7 parts per million. This amount of methane would be hardly detectable by a telescope on a planet around another star. But when Segura used that production of methane in the simulated earths around M dwarfs, this compound built up in the atmosphere, thus the model crashed. Antígona learned how to solve the problem, and with her collaborators showed that the particular UV emission of these M dwarfs changed the production of a compound that was responsible for methane destruction. As a result CH_4 along with other atmospheric gases produced by life were more abundant and therefore more likely to be detected. It this way methane can be used as a potential biosignature (Segura et al. 2005).

We are interested on M dwarfs because they are about 75% of the stars in our galaxy, and most of the potentially habitable planets are around M dwarfs. A habitable planet is one that is capable of hosting life, although science has not yet a definition for life. Not having a definition may seem like a problem for identifying habitable planets, but scientists are experts in solving problems. What we do is to take the most general characteristics of life in the only inhabited planet we know:

Earth. Although living beings on Earth may look very different from each other – think for example in a fungus, a bacterium, a shark, and a tree – looking closer, life is made with a few elements: hydrogen, carbon, oxygen, nitrogen, sodium, magnesium, potassium, calcium, phosphorus, and sulfur (see Chap. 1). The first four in the list are highly common in the universe, and carbon is the element that binds the rest of the elements to form the molecules that build living beings. For carbon molecules to find each other and react to create new structures, they require a substance where they can move. On Earth, this substance is liquid water. Then, the generalization of Earth's life is carbon and liquid water. The distance from the star where a rocky planet with water and atmosphere can have liquid water on its surface is called the Habitable Zone (HZ) (Chap. 1).

After the 2005 work by Segura and collaborators demonstrated the importance of UV to predict the biosignatures for planets around M dwarfs, there was a renewed interest on these stars (Scalo et al. 2007; Tarter et al. 2007; Billings 2011; Shields et al. 2016) strengthened by the detection of several planets in the habitable zone of M dwarfs including our closest neighbor, Proxima Centauri (Anglada-Escudé et al. 2016). Because the HZ of M dwarfs is 10–100 times closer than the Sun's habitable zone, it was important to measure how much high-energy radiation may reach the surface of potentially habitable planets around them. So, we needed to measure the UV of non-active stars (Hawley 2005). Observations with the *Hubble Space Telescope* finally showed us that UV emission of "non-active" M dwarfs was higher than we expected (Walkowicz et al. 2008), and later the MUSCLES treasury survey started gathering a catalog for the emission (from infrared to X-rays) of M dwarfs with exoplanets detected (France 2016). During the last decade, the astrobiology community has been discussing if planets would be habitable and life can emerge and evolve in those harsh conditions.

M dwarfs have flares, sudden discharges of energy that increase the UV emission of active M dwarfs 10–100 times during a few hours every day. This UV could potentially destroy the ozone (O_3) in a planetary atmosphere like the present Earth, representing a problem for planetary habitability. Segura worked on this problem when she moved to Pasadena, CA, hired by JPL to collaborate in the Virtual Planetary Laboratory lead by Dr. Victoria Meadows. This time, she modified the code to include the flux of a flare from AD Leonis finding that actually, their simulated earths did not receive much more UV energy than Earth in the past when life emerged, and there was no ozone in the atmosphere (Segura et al. 2010). By the end of 2006, she was hired at the UNAM to continue her work on M dwarfs, now as a full-time professor.

Since Antígona moved to Mexico, along with students and collaborators, they have been exploring some of the problems for habitability and life detection on planets around M dwarfs. For example, with Jesús Zendejas and Alejandro Raga, they explore the atmospheric loss of planets produced by the interaction with the M dwarfs stellar wind. Without the protection of a magnetic field, planets around active M dwarfs will be left with no atmosphere in one billion years (Zendejas et al. 2010). With Lisa Kaltenegger, they made predictions about the atmosphere needed for a planet called GJ 581 d to be potentially habitable (Kaltenegger et al. 2011).

This exoplanet was the first found in the habitable zone; although it was later demonstrated that the signal was produced by stellar activity and not by an exoplanet (Robertson et al. 2014), the community learned about how the different assumptions and models used changed the predictions for the abundance of CO_2 that this exoplanet may need to be habitable (Domagal-Goldman and Segura 2013).

High-energy radiation is not necessarily a hazard for life. Our planet was not protected from UV irradiation during its first 2 billion years and anyway life emerged and thrived. Then, Segura leads a group of students who is trying to cover all the possibilities. Arturo Miranda and David Ramos explored the chemistry of atmospheres with more CO_2 and less O_2 under the UV light of M dwarfs to understand how biosignatures may change for these cases. Marion Zulema Armas is working on the possibility that the UV light was the driver for building the starting materials for life in planets in the HZ of M dwarfs. Laura Ribeiro Do Amaral is using a model called VPLanet to find out if an atmosphere will survive to the X-rays from the frequent and energetic flares of an M dwarf. For now, we cannot detect atmospheres of exoplanets around M dwarfs; new instruments are being planned for doing the task. This research will help to choose the best targets to observe with those instruments and to interpret observations. We expect to discover habitable planets and maybe find the signatures of life in one of them to tell us we are not alone in the universe. A happy thought even if our neighbors are just gas-producing bacteria.

References

Alba-Aldave LA et al (2009) Ca-Poor pyroxene Raman characteristics in H ordinary chondrites. In: AIP conference proceedings. AIP, Mainz, pp 161–167

Anglada-Escudé G et al (2016) A terrestrial planet candidate in a temperate orbit around Proxima Centauri. Nature 536:437–440

Avendaño RE et al (2015) Growth of *Bacillus pumilus* and *Halomonas halodurans* in sulfates: prospects for life on Europa. B Soc Geol Mex 67(3):367–375

Bailey ME, Markham DJ, Massai S, Scriven JE (1995) The 1930 "Brazilian Tunguska" event. Observatory 115:250–253

Bakos GA et al (2002) System description and first light curves of the Hungarian automated telescope, an autonomous observatory for variability search. Publ Astron Soc Pac 114:974–987

Barge LM, White LM (2017) Experimentally testing hydrothermal vent origin of life on Enceladus and other icy/ocean worlds. Astrobiology 17(9):820–833

Barragán O et al (2018) K2-141 b. A 5-M_Earth super-Earth transiting a K7 V star every 6.7 h. Astron Astrophys 612:A95

Barros SCC et al (2011) WASP-38b: a transiting exoplanet in an eccentric, 6.87d period orbit. Astron Astrophys 525:A54

Béghin C et al (2009) New insights on Titan's plasma-driven Schumann resonance inferred from Huygens and Cassini data. Planet Space Sci 57(14–15):1872–1888

Beltrán Y et al (2012) N_2 fixation rates and associated diversity (*nifH*) of microbialite and mat-forming consortia from different aquatic environments in Mexico. Aquat Microb Ecol 67:15–24

Biemann K (2007) On the ability of the Viking gas chromatograph–mass spectrometer to detect organic matter. Proc Natl Acad Sci U S A 104(25):10310–10313

Biemann K, Bada JL (2011) Comment on "Reanalysis of the Viking results suggests perchlorate and organics at midlatitudes on Mars" by Rafael Navarro-González et al. J Geophys Res Planet 116(E12):E12001

Billings L (2011) Astronomy: exoplanets on the cheap. Nature News 470(7332):27–29

Borucki WJ et al (2010) Kepler planet-detection Mission: introduction and first results. Science 327:977

Brito EM et al (2014) Microbial diversity in Los Azufres geothermal field (Michoacán, Mexico) and isolation of representative sulfate and sulfur reducers. Extremophiles 18:385–398

Brito EM et al (2019) The bacterial diversity on steam vents from Paricutín and Sapichu volcanoes. Extremophiles 23:249–263

Brown ME et al (2002) Direct detection of variable tropospheric clouds near Titan's south pole. Nature 420:795–797

Centeno CM et al (2012) Microbialite genetic diversity and composition relate to environmental variables. FEMS Microbiol Ecol 82:724–735

Ceplecha Z et al (1993) Atmospheric fragmentation of meteoroids. Astron Astrophys 279:615–626

Cervantes K et al (2017) Tequisquiapan. Meteoritical Bull 103:173

Cervantes-de la Cruz KE et al (2010) Termometría de dos piroxenos en condros de la condrita ordinaria Nuevo Mercurio H5, México. Rev Mex Cienc Geol 27(1):134–147

Cervantes-de la Cruz KE et al (2015) Experimental chondrules by melting samples of olivine, clays and carbon with a CO_2 laser. Bol Soc Geol Mex 67:401–412

Chacón E et al (2018) Biohermal thrombolites of the crater lake Rincón de Parangeo in Central México. J S Am Earth Sci 85:236–249

Chen LX et al (2018) Metabolic versatility of small archaea Micrarchaeota and Parvarchaeota. ISME J 12:756–775

Coates AJ et al (2009) Heavy negative ions in Titan's ionosphere: altitude and latitude dependence. Planet Space Sci 57(14–15):1866–1871

Coll P et al (1999) Experimental laboratory simulation of Titan's atmosphere: aerosols and gas phase. Planet Space Sci 47(10/11):1331–1340

Connelly JN et al (2012) The absolute chronology and thermal processing of solids in the solar protoplanetary disk. Science 338:651–655

Cordero G, Mendoza B (2004) Evidence for the origin of ridges on Europa by means of photoclinometric data from E4 Galileo orbit. Geofis Int 43(2):301–306

Cordero G, Poveda A (2011) Curuça 1930: A probable mini-Tunguska? Planet Space Sci 59(1):10–16

Cordero G et al (2011) The bolide of February 10, 2010: observations in Hidalgo and Puebla, Mexico. Geofis Int 50(1):77–84

Cordero-Tercero G et al (2016) The Mexican meteor network: a preliminary proposal. Geofis Int 55(1):69–71

Corona-Chávez P et al (2018) Petrology, phase equilibria modelling, noble gas chronology and thermal constraints of the El Pozo L5 meteorite. Geochemistry 78:248–253

Couradeau E et al (2011) Prokaryotic and eukaryotic community structure in field and cultured Microbialites from the alkaline Lake Alchichica (Mexico). PLoS One 6:e28767

Coustenis A (2014) Titan. In: Spohn T, Breuer D, Johnson TV (eds) Encyclopedia of the solar system, 3rd edn. Elsevier, Amsterdam, pp 831–849

Curiel S et al (2011) A fourth planet orbiting Upsilon Andromedae. Astron Astrophys 525:A78

De Anda V et al (2018) Understanding the mechanisms behind the response to environmental perturbation in microbial mats: a metagenomic-network based approach. Front Microbiol 9:2606

Del Castillo A (1864) Descripción de la masa de hierro meteórico de Yanhuitlán recientemente traída a esta capital. Boletín de la Sociedad Mexicana de Geografía y Estadística 10:661–665

Del Castillo A (1889) Catalogue descriptif des Météorites (Fers et pierres Météoriques) du Mexique, Paris

Del Castillo A (1893) Carta de los meteoritos de México o regiones de la república en que han caído fierro y piedras meteóricas Escala. 1:10, 000, 000. E. Moreau

Demory B-O, Seager S (2011) Lack of inflated radii for Kepler giant planet candidates receiving modest stellar irradiation. Astrophys J Suppl S 197(1):12

Dick SJ (2013) The twentieth century history of the extraterrestrial life debate: major themes and lessons learned. In: Vakoch DA (ed) Astrobiology, history, and society: life beyond earth and the impact of discovery. Springer, Berlin/Heidelberg, pp 133–173

Dohnanyi JS (1969) Collisional model of asteroids and their debris. J Geophys Res 74(10):2531–2554

Domagal-Goldman SD, Segura A (2013) Exoplanet climates. In: Mackwell SJ et al (eds) Comparative climatology of terrestrial planets. University of Arizona Press, Tucson, pp 121–135

Doyle LR et al (2011) Kepler-16: a transiting circumbinary planet. Science 333:1602

Eigenbrode JL et al (2018) Organic matter preserved in 3-billion-year-old mudstones at Gale crater, Mars. Science 360(6393):1096–1101

Falchi F et al (2016) The new world atlas of artificial night Sky brightness. Sci Adv 2(6):e1600377

Farinella P et al (2001) Probable Asteroidal origin of the Tunguska cosmic body. Astron Astrophys 377:1081–1097

Flores-Gutiérrez D et al (2010a) Scanning electron microscopy characterization of iron, nickel and sulfur in chondrules from the Allende meteorite–further evidence for between-chondrules major compositional differences. Rev Mex Cienc Geol 27:338–346

Flores-Gutiérrez D et al (2010b) Micromagnetic and microstructural analyses in chondrules of the Allende meteorite. Rev Mex Cienc Geol 27:162–174

Fortes AD (2000) Exobiological implications of a possible ammonia-water ocean inside Titan. Icarus 146:444–452

France K (2016) The MUSCLES treasury survey. I. Motivation and overview. Astrophys J 820(2):89

Freissinet C et al (2015) Organic molecules in the Sheepbed Mudstone, Gale Crater, Mars. J Geophys Res Planet 120(3):495–514

Fulton BJ, Petigura EA (2018) The California-Kepler survey. VII. Precise planet radii leveraging Gaia DR2 reveal the stellar mass dependence of the planet radius gap. Astron J 156(6):264

García Fajardo R (2017) Diseño modular para una estación de monitoreo de bólidos. Facultad de Ingeniería, UNAM. Bachelor thesis

García-Martínez JL, Ortega-Gutiérrez F (2008) Streamlets within meteoroid streams on NEA orbits. Geofis Int 47:251–255

Gibbard SG et al (2004) Speckle imaging of Titan at 2 microns: surface albedo, haze optical depth, and tropospheric clouds 1996–1998. Icarus 169:429–439

Gillon M et al (2016) Temperate Earth-sized planets transiting a nearby ultracool dwarf star. Nature 533:221–224

Gillon M et al (2017) Seven temperate terrestrial planets around the nearby ultracool dwarf star TRAPPIST-1. Nature 54:456–460

Gischler E et al (2008) Giant holocene freshwater microbialites, Laguna Bacalar, Quintana Roo, Mexico. Sedimentology 55:1293–1309

Glein CR et al (2015) The pH of Enceladus' ocean. Geochim Cosmochim Acta 162:202–219

Goffredi SK et al (2017) Hydrothermal vent fields discovered in the southern Gulf of California clarify role of habitat in augmenting regional diversity. Proc Biol Sci 284(1859):20170817

Gómez Maqueo Chew Y et al (2013a) Discovery of WASP-65b and WASP-75b: two hot Jupiters without highly inflated radii. Astron Astrophys 559:A36

Gómez Maqueo Chew Y et al (2013b) The homogeneous study of transiting systems (HoSTS). I. the pilot study of WASP-13. Astrophys J 768:79

González-Medina K (2019) Estudio Petrológico de la Meteorita Condrítica Aldama b; H5; S2; W3. Bachelor Dissertation, Universidad Autónoma de San Luis Potosí

González JJ et al (2018) Recent developments at the OAN-SPM. In: Ground-based and airborne telescopes VII. SPIE astronomical telescopes + instrumentation, Austin, Texas, United States, 2018. Society of Photo-Optical Instrumentation Engineers (SPIE) conference series, vol 10700. SPIE, Bellingham, p 107005F

Griffith CA, Owen T, Miller GA, Geballe T (1998) Transient clouds in Titan's lower atmosphere. Nature 395(6702):575–578

Griffith CA, Hall JL, Geballe TR (2000) Detection of daily clouds on Titan. Science 290:509–513

Hernández-Bernal MS, Solé J (2010) Edades K-Ar y Pb-Pb de condros individuales de condritas ordinarias mexicanas como trazadores de eventos de impacto prolongadas. Rev Mex Cienc Geol 27:123–133

Hand KP, Carlson RW (2015) Europa's surface color suggests an ocean rich with sodium chloride. Geophys Res Lett 42(9):3174–3178

Hand KP, Chyba CF (2007) Empirical constraints on the salinity of the Europan Ocean and implications for a thin ice shell. Icarus 189(2):424–438

Haro JC (1931) Las meteoritas mexicanas. generalidades sobre meteoritas y catálogo descriptivo de las meteoritas mexicanas, Especial. Universidad Nacional Autónoma de México

Harris JH et al (2012) Phylogenetic stratigraphy in the Guerrero Negro hypersaline microbial mat. ISME J 7:60–60

Hawley S (2005) Characterizing the near-UV environment of M dwarfs: implications for extrasolar planetary searches and astrobiology. Hubble Space Telescope Proposal

Hendrix AR et al (2018) The NASA Roadmap to Ocean Worlds. Astrobiology 19(1):1–27

Hernández-Zavala ME (2016) Caracterización hidrogeoquímica y diversidad microbiana en ambientes extremos, casos: Villa Juárez, San Luis Potosí y Rincón de Parangueo, Guanajuato. IPICYT, Master thesis

Herrera AL (1919) Some studies in plasmogenesis. J Lab Clin Med 4(8):479–483

Hibbitts CA et al (2019) Evidence for the presence of sulfates on Europa. Icarus 326:37–47

Hills JG, Goda MP (1993) The fragmentation of small asteroids in the atmosphere. Astron J 105(3):1114–1144

Howard AW et al (2010) The occurrence and mass distribution of close-in super-Earths, Neptunes, and Jupiters. Science 330:653–655

Israel G et al (2005) Complex organic matter in Titan's atmospheric aerosols from in situ pyrolysis and analysis. Nature 438(7069):796–799

Johnson PV et al (2019) Insights into Europa's ocean composition derived from its surface expression. Icarus 321:857–865

Jones DS et al (2016) Biogeography of sulfur-oxidizing *Acidithiobacillus* populations in extremely acidic cave biofilms. ISME J 10:2879

Kaltenegger L et al (2011) Model spectra of the first potentially habitable super-Earth—Gl581d. Astrophys J 733(1):35

King EA et al (1969) Meteorite Fail at Pueblito de Allende, Chihuahua, Mexico: Preliminary Information. Science 163:928

Krajick K (2007) Robot seeks new life—and new funding: in the Abyss of Zacatón. Science 315:322–324

Kulik LA (1931) The Brazilian twin of the Tunguska meteorite. Priroda I Ljudi 13(14):6–11

L'Osservatore Romano (Informazioni Fides) (1931) La caduta di tre bolide alle Amazzoni 50. 1 March: 5

Ledesma-Mateos I, Barahona A (2003) The institutionalization of biology in Mexico in the early 20th century. The conflict between Alfonso Luis Herrera (1868–1942) and Isaac Ochoterena (1885–1950). J Hist Biol 36(2):285–307

Ley RE et al (2006) Unexpected diversity and complexity of the Guerrero Negro hypersaline microbial mat. Appl Environ Microbiol 72:3685–3695

López-García K (2019) Caracterización de una meteorita metálica de la localidad de Real de Catorce, S.L.P. Bachelor Dissertation, Universidad Nacional Autónoma de México

Lorenz RD et al (2008) Titan's rotating reveals an internal ocean and changing zonal winds. Science 319(5870):1649–1651

Lorenz RD et al (2011) Analog environments for a Europa lander mission. Adv Space Res 48(4):689–696

Lovelock JE (1965) A physical basis for life detection experiments. Nature 207:568–570

Lowell P (1908) The habitability of Mars. Nature 77(2003):461–461

Martin W et al (2008) Hydrothermal vents and the origin of life. Nat Rev Microbiol 6(11):805

Mason B (1975) Allende meteorite. Cosmochemistry's Rosetta Stone. Acc Chem Res 8:217–224

Mayor M, Queloz D (1995) A Jupiter-mass companion to a solar-type star. Nature 378:355–359

Medrano-Santillana M et al (2017) Bacterial diversity in fumarole environments of the Paricutín volcano, Michoacán (Mexico). Extremophiles 21:499–511

Melosh HJ (1989) Impact cratering. A geological process. Oxford University Press, Oxford

Melosh HJ (2011) Planetary surface processes. Cambridge University Press, Cambridge, pp 7–8

Michel P, Morbidelli A (2013) Population of impactors and the impact cratering rate in the inner solar system. In: Osinski GR, Pierazzo E (eds) Impact cratering, processes and products. Wiley-Blackwell, Hoboken, pp 21–31

Miller SL (1953) A production of amino acids under possible primitive Earth conditions. Science 117:528–529

Ming DW et al (2014) Volatile and organic compositions of sedimentary rocks in Yellowknife Bay, Gale Crater, Mars. Science 343(6169):1245267

Montoya-Pérez M (2016) Caracterización morfológica y mineralógica de condros barrados de olivino presentes en la condrita Allende. Bachelor dissertation, Universidad Nacional Autónoma de México

Montoya-Pérez M (2019) Clasificación de la condrita NWA-M por medio de métodos no destructivo. Master dissertation, Universidad Nacional Autónoma de México

Montoya L et al (2002) Hydrocarbon production under hydrothermal Europan conditions and its implications to the maintenance of life. In: ISSOL '02, Oaxaca, Mexico, June 30 to July 5, 2002. Origins Life Evol Bisoph 32:405–546

Montoya L et al (2011) The sulfate-rich and extreme saline sediment of the ephemeral Tirez lagoon: a biotope for acetoclastic sulfate-reducing bacteria and hydrogenotrophic methanogenic archaea. Int J Microbiol 2011:753758

Montoya L et al (2017) Los ambientes extremos en Rincón de Parangueo, Gto. y Villa Juárez, SLP, y el efecto del pH sobre la diversidad microbiana. In: X Congreso Nacional de Astrobiología, Monterrey, Mexico, 7–10 March 2017, SOMA

Navarro KF et al (2010) Chemical signatures of life in modern stromatolites from Lake Alchichica, Mexico. Applications for the search of life on Mars. In: 38th COSPAR scientific assembly, 38:13

Navarro-González R (1998) In memoriam: Cyril Andrew Ponnamperuma, 1923–1994. Orig Life Evol Bisoph 28(2):105–108

Navarro-González R, Ramírez SI (1997) Corona discharge of Titan's troposphere. Adv Space Res 19(7):1121–1133

Navarro-González R, McKay C (2011) Reply to comment by Biemman and Bada on "Reanalysis of the Viking results suggests prechlorate and organics at midlatitudes on Mars. J Geophys Res Planets 116(E12):E12002

Navarro-González R et al (2006) The limitations on organic detection in Mars-like soils by thermal volatilization–gas chromatography–MS and their implications for the Viking results. Proc Natl Acad Sci U S A 103(44):16089–16094

Navarro-González R et al (2010) Reanalysis of the Viking results suggests perchlorate and organics at midlatitudes on Mars. J Geophys Res Planet 115:E12

Negrón-Mendoza A (1995) Alfonso L. Herrera: a Mexican pioneer in the study of chemical evolution. J Biol Phys 20(1):11–15

Nimmo F, Pappalardo RT (2016) Ocean worlds in the outer solar system. J Geophys Res Planets 121(8):1378–1399

O'Keefe JD, Ahrens TJ (1977) Meteorite impact Ejecta: dependence of mass and energy lost on planetary escape velocity. Science 198:1249–1251

Oparin AI (1924) The origin of life, transl. A. Synge, in J. D. Bernal (1967), The origin of life. Oxford University Press, London, pp 197–234

Osinski GR, Pierazzo E (Eds) (2013) Impact cratering, processes and products, Hoboken: Wiley-Blackwell 333

Palme H et al (1981) The Acapulco meteorite: chemistry, mineralogy and irradiation effects. Geochim Cosmochim Acc 45:727–752

Passey QR, Melosh HJ (1980) Effects of atmospheric breakup on crater field formation. Icarus 42:211–233

Pepe F et al (2004) The HARPS search for southern extrasolar planets. I. HD 330075 b: a new hot Jupiter. Astron Astrophys 423:385–389

Petigura EA, Howard AW, Marcy GW (2013) Prevalence of Earth-size planets orbiting Sun-like stars. Proc Natl Acad Sci U S A 110(48):19273–19278

Pollacco DL et al (2006) The WASP project and the SuperWASP cameras. Publ Astron Soc Pac 118:1407–1418

Postberg F et al (2011) A salt-water reservoir as the source of a compositionally stratified plume on Enceladus. Nature 474(7353):620–622

Postberg F et al (2018) Macromolecular organic compounds from the depths of Enceladus. Nature 558(7711):564

Potter EG et al (2009) Isotopic composition of methane and inferred methanogenic substrates along a salinity gradient in a hypersaline microbial mat system. Astrobiology 9:383–390

Poveda A, Cordero G (2008) Chicxulubites: a new class of meteorites? Geofis Int 47(3):167–172

Quirrenbach A et al (2014) CARMENES instrument overview. In: Ground-based and Airborne Instrumentation for Astronomy. Proceedings of the SPIE 9147:91471F

Ramírez Cabañas AK (2017) Procesos internos y externos asociados al criovulcanismo de Encélado. UNAM, Mexico, thesis

Ramírez SI et al (2001) Possible contribution of different energy sources to the production of organics in Titan's atmosphere. Adv Space Res 27:261–270

Ramírez SI et al (2004) Chemical characterization of aerosols in simulated planetary atmospheres. Titan's aerosol analogues. In: Seckbach J, Chela-Flores JTO, Raulin F (eds) Life in the Universe. From the Miller experiment to the search for life on other worlds, vol 7. Cellular origin, life in extreme habitats and astrobiology, vol 7. Kluwer Academic Publishers, Dordrecht, pp 281–285

Ramírez SI et al (2010) The fate or aerosols on the surface of Titan. Faraday Discuss 147:419–427

Reyes-Salas AM et al (2010) Petrography and mineral chemistry of Escalón meteorite, an H4 chondrite, México. Rev Mex Cienc Geol 27:148–161

Reza R et al (2004) The event near the Curuça river. August 67th Annual Meteoritical Society Meeting. Meteoritics Planet Sci Suppl 39:A30

Ricker GR et al (2014) Transiting Exoplanet Survey Satellite (TESS). In: Space telescopes and instrumentation 2014: optical, infrared, and millimeter wave, vol 9143 of Proceedings of the SPIE, p 914320

Robertson P et al (2014) Stellar activity masquerading as planets in the habitable zone of the M dwarf Gliese 581. Science 345(6195):440–444

Roe HG et al (2002) Titan's clouds from Genini and Keck adaptive optics imaging. Astrophys J 581:1399–1406

Roth L et al (2014) Transient water vapor at Europa's South Pole. Science 343(6167):171

Rubin AE, Grossman JN (2010) Meteorite and meteoroid: new comprehensive definitions. Meteorit Planet Sci 45(1):114–122

Ruiz J et al (2007) Thermal diapirism and the habitability of the icy shell of Europa. Origins Life Evol Biosph 37(3):287–295

Ruvinovich-Kogan R (1992) Los meteoritos y glosario de meteorítica. Bol Mineral 5:96–113

Sahl JW et al (2010) Novel microbial diversity retrieved by autonomous robotic exploration of the world's deepest vertical phreatic sinkhole. Astrobiology 10:201–213

Sánchez Rubio G (1978) Meteoritical bulletin, no. 55. Meteoritics 13:327–328

Sánchez Rubio G et al (2001) Las Meteoritas de México. Instituto de Geología UNAM, Instituto de Geofísica, UNAM, México, D.F

Sánchez-Sánchez J et al (2019) Extant microbial communities in the partially desiccated Rincon de Parangueo maar crater lake in Mexico FEMS. Microbiol Ecol 95:fiz051

Scalo J et al (2007) M stars as targets for terrestrial exoplanet. Astrobiology 7(1):85–166

Schneider J et al (2011) Defining and cataloging exoplanets: the exoplanet.eu database. Astron Astrophys 532:A79

Scipioni F et al (2017) Deciphering sub-micron ice particles on Enceladus surface. Icarus 290:183–200

SEAN (1978) Acapulco fall. Sci Event Alert Netw Bull 3:10

Segura A (2018) Star-planet interactions and habitability: radiative effects. In: Deeg HJ, Belmonte JA (eds) Handbook of exoplanets. Springer International Publishing, Cham, pp 1–23. https://doi.org/10.1007/978-3-319-30648-3_73-1

Segura A et al (2003) Ozone concentrations and ultraviolet fluxes on Earth-like planets around other stars. Astrobiology 3(4):689–708

Segura A et al (2005) Biosignatures from Earth-like planets around M dwarfs. Astrobiology 5(6):706–725

Segura A et al (2010) The effect of a strong stellar flare on the atmospheric chemistry of an Earth-like planet orbiting an M dwarf. Astrobiology 10(7):751–771

Sekine Y et al (2015) High-temperature water–rock interactions and hydrothermal environments in the chondrite-like core of Enceladus. Nat Commun 6(1):8604

Shields AL et al (2016) The habitability of planets orbiting M-dwarf stars. Phys Rep 663:1–38

Siefert J (2011) How a real housewife became an astrobiologist. Astrobiology 11:193–195

Siverd RJ et al (2012) KELT-1b: a strongly irradiated, highly inflated, short period, 27 Jupiter-mass companion transiting a Mid-F Star. Astrophys J 761:123

Soriano-López CA et al (2017) There is water and food on Enceladus satellite, what is next? In: UAEM (ed) International symposium "extreme ecosystems and extremophile organisms: biodiversity, physiology, biochemistry and biotechnology, Cuernavaca, Mexico, 18–20 September 2017, p 92

Sotin C et al (2010) Titan's interior structure. In: Brown R, Lebreton JP, Waite JH (eds) Titan from Cassini-Huygens, 1st edn. Springer, Dordrecht, pp 61–73

Souza V et al (2012) The Cuatro Ciénegas Basin in Coahuila, Mexico: an astrobiological precambrian park. Astrobiology 12:641–647

Squyres SW et al (1983) Liquid water and active resurfacing on Europa. Nature 301(5897):225–226

Talens GJJ et al (2017) The Multi-site All-Sky CAmeRA (MASCARA). Finding transiting exoplanets around bright (m_V < 8) stars. Astron Astrophys 601:A11

Tapia-Torres Y et al (2016) How to live with phosphorus scarcity in soil and sediment: lessons from bacteria. Appl Environ Microbiol 82:4652–4662

Tarter JC et al (2007) A reappraisal of the habitability of planets around M dwarf stars. Astrobiology 7(1):30–65

Taubner R-S et al (2018) Biological methane production under putative Enceladus-like conditions. Nat Commun 9(1):748

Tazaz AM et al (2013) Redefining the isotopic boundaries of biogenic methane: methane from endoevaporites. Icarus 224(2):268–275

Teske A et al (2003) Genomic markers of ancient anaerobic microbial pathways: sulfate reduction, methanogenesis, and methane oxidation. Biol Bull 204:186–191

Tobie G et al (2005) Titan's internal structure inferred from a coupled thermal-orbital model. Icarus 175(2):496–502

Tobler M et al (2008) Toxic hydrogen sulfide and dark caves: phenotypic and genetic divergence across two abiotic environmental gradients in Poecilia mexicana. Evolution 62:2643–2659

Tomasko MG, West RA (2010) Aerosols in Titan's atmosphere. In: Brown R, Lebreton JP, Waite JH (eds) Titan from Cassini-Huygens, 1st edn. Springer, Dordrecht, pp 297–321

Trumbo SK et al (2019) Sodium chloride on the surface of Europa. Sci Adv 5(6):eaaw7123

Urrutia-Fucugauchi J et al (2014) Meteorite paleomagnetism – from magnetic domains to planetary fields and core dynamos. Geofis Int 53:343–363

Valdespino-Castillo PM et al (2018) Exploring biogeochemistry and microbial diversity of extant microbialites in Mexico and Cuba. Front Microbiol 9:510

Valdivieso-Ojeda JA et al (2014) High enrichment of molybdenum in hypersaline microbial mats of Guerrero Negro, Baja California Sur, Mexico. Chem Geol 363:341–354

Vance SD (2018) The habitability of Icy Ocean worlds in the solar system. In: Deeg HJ, Belmonte JA (eds) Handbook of exoplanets. Springer International Publishing, Cham, pp 2855–2877

Vance SD et al (2018) Geophysical investigations of habitability in ice-covered Ocean worlds. J Geophys Res Planet 123(1):180–205

Vasilyev N, Andreev G (1989) The Brazilian twin of the Tunguska meteorite: myth or reality? WGN. J Int Meteor Organ 17(6):247–248

Vickery A (1987) Variation in ejecta size with ejection velocity. Geophys Res Lett 14:726–729

Vogel MB et al (2009) The role of biofilms in the sedimentology of actively forming gypsum deposits at Guerrero Negro, Mexico. Astrobiology 9:875–893

Wahlund JE et al (2009) On the amount of heavy molecular ions in Titan's ionosphere. Planet Space Sci 57(14–15):1857–1865

Waite JH et al (2007) The process of tholin formation in Titan's upper atmosphere. Science 316:870–875

Walkowicz LM et al (2008) Characterizing the near-UV environment of M dwarfs. Astrophys J 677(1):593

Wolszczan A (1994) Confirmation of Earth-mass planets orbiting the millisecond pulsar PSR B1257+12. Science 264:538–542

Wolszczan A, Frail DA (1992) A planetary system around the millisecond pulsar PSR1257 + 12. Nature 355:145–147

Zendejas J et al (2010) Atmospheric mass loss by stellar wind from planets around main sequence M stars. Icarus 210(2):539–544

Zolotov MY (2007) An oceanic composition on early and today's Enceladus. Geophys Res Lett 34:L23203. https://doi.org/10.1029/2007GL031234

Zolotov MY, Shock EL (2003) Energy for biologic sulfate reduction in a hydrothermally formed ocean on Europa. J Geophys Res Planet 108(E4):5022. https://doi.org/10.1029/2002JE001966

Chapter 3
Was LUCA a Hyperthermophilic Prokaryote? The Impact-Bottleneck Hypothesis Revisited

Gilberto P. Morales and Luis Delaye

Abstract In the *Origin of Species*, Darwin wrote "The affinities of all the beings of the same class have sometimes been represented by a great tree. I believe this simile largely speaks the truth." Modern comparative genomics has revealed that the intuition of Darwin was correct. A set of highly conserved genes and cellular functions indicate that all life on Earth is related by common ancestry. These genes were inherited from the last universal common ancestor or LUCA. The functions coded by these genes suggest that LUCA was a rather complex cell already endowed with a genetic code and a protein translation apparatus. One of the questions regarding the nature of LUCA is whether it was a hyperthermophile. Here, we review recent evidence derived from the molecular fossil record on the temperature preferences of LUCA. We suggest that current evidence on the nature of LUCA and its immediate predecessors are compatible with the impact-bottleneck hypothesis – the proposal that during the early evolution of life, a meteoritic impact eliminated all life on Earth except for prokaryotes capable of living at high temperatures. If our interpretation of the data is correct, it would indicate that early life was resilient to the rough environmental conditions of the Archean, a relevant result from the point of view of astrobiology because it would exemplify the persistence of life in harsh environments.

3.1 Early Life as Revealed by Molecular Sequences

The appearance of early life on the primitive Earth was preceded by a diversity of astronomical and geochemical events that together set the conditions for the origin of life. From the prebiotic point of view, the timing and the specific conditions that led to the origin of organic molecules and the assembly and evolution of these into

G. P. Morales · L. Delaye (✉)
Departamento de Ingeniería Genética, CINVESTAV, Irapuato, Guanajuato, Mexico
e-mail: luis.delaye@cinvestav.mx

© Springer Nature Switzerland AG 2020
V. Souza et al. (eds.), *Astrobiology and Cuatro Ciénegas Basin as an Analog of Early Earth*, Cuatro Ciénegas Basin: An Endangered Hyperdiverse Oasis,
https://doi.org/10.1007/978-3-030-46087-7_3

the first cellular systems are still active areas of research (Patel et al. 2015). However, from a biological perspective, there is strong evidence indicating that extant cells were preceded by simpler systems where RNA performed both catalytic and hereditary functions (Becerra et al. 2007). These primitive systems based on RNA first evolved protein synthesis leading to RNA/protein cells and then DNA synthesis (Freeland et al. 1999).

As extant cells use DNA as their hereditary material, parsimony indicates that the last universal common ancestor (or simply LUCA) had a DNA-based genome. However, our ability to infer the gene content and therefore the nature of LUCA is largely determined by the structure and the evolutionary process that generated the universal Tree of Life (ToL). By using the information contained in the small subunit (SSU) of rRNA, Woese and Fox proposed that all life on Earth can be divided in three main lines of descent; these are Archaeobacteria (today Archaea), Bacteria and the nucleo-cytoplasm component of Eukarya (Woese and Fox 1977a). Incidentally, Woese and Fox also provided evidence of the bacterial origin of organelles, thus supporting the hypothesis proposed by Lynn Margulis for the origin of mitochondria and plastids (Sagan 1967). In the same year, Woese and Fox also proposed that all three cellular domains diverged from an ancestral entity they named *progenote*. They defined the *progenote* as a primitive entity "to recognize the possibility that it had not yet completed evolving the link between genotype and phenotype" (Woese and Fox 1977b). Several years later, Woese and coworkers formally proposed that all life on Earth should be classified on the basis of the SSU rRNA phylogeny into Archaea, Bacteria, and Eukarya and that the root of such ToL was located in the bacterial branch (Woese et al. 1990). Although biologists have largely adopted this phylogenetic scheme to represent the evolutionary relationships between all living beings, there have been attempts to improve this view of the ToL (Fig. 3.1).

Early criticisms to the three-domain view of the ToL suggested that it was inaccurate to represent the evolution of genomes by using a single gene encoding the small subunit rRNA (SSU rRNA) molecule because genes among prokaryotes were often transferred between lineages. Accordingly, the evolution of life on Earth was better represented by a net rather than by a bifurcating tree (Doolittle 1999). For instance, Fournier et al. (2009) used the metaphor of the "coral of life" to better depict the ToL due to the highly connected network of the coral dead layers that were analogous to horizontal gene transfer (HGT) events. In a similar fashion, Koonin et al. (2009) used the metaphor of a forest of life to summarize the phylogenetic history of multiple genes along evolution.

Other modifications to the scheme proposed by Woese and Fox concerned the position of eukaryotes in the ToL. Early on, James Lake proposed that eukaryotes branched from a group of Archaea named by him as Eocytes (Lake et al. 1984). The Eocytes were later identified as Crenarchaeota. Interestingly, a modified version of this hypothesis has received support from recent phylogenetic analyses. These analyses suggest that Eukarya evolved from within the Archaea. Specifically, that Eukarya are closely related to a lineage of Archaea known as TACK: Thaumarchaeota, Aigarchaeota, Crenarchaeota and Korarchaeota (Spang et al. 2015).

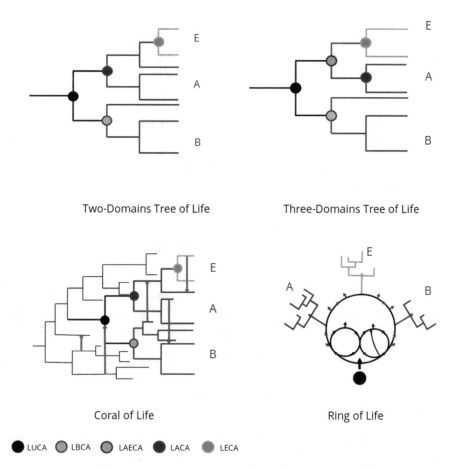

Fig. 3.1 Different proposals of the Tree of Life (ToL). Circles represent common ancestors of Archaea (A), Bacteria (B), and Eukarya (E) domains. *LUCA* last universal common ancestor, *LBCA* last bacterial common ancestor, *LAECA* last archaeal-eukaryal common ancestor, *LACA* last archaeal common ancestor, *LECA* last eukaryotic common ancestor. The arrows represent the horizontal gene transfer process

An alternative version of the origin of eukaryotes (and the treelike structure) was proposed by Rivera and Lake (2004) where the ToL is substituted by a ring of life. In this scheme, Eukarya originated by a fusion between a Gram-negative prokaryote and an Eocyte, thus giving rise to a ringlike structure. Nevertheless, the ring of life structure has been criticized for being biologically unsound (Forterre 2015). More recently, there have been attempts to add a fourth domain of life to the ToL conformed only by viruses (Boyer et al. 2010). However, the viral fourth domain of life hypothesis has not resisted more careful phylogenetic analyses (Moreira and López-García 2015).

As mentioned above, it is crucial to know the structure of the ToL to be able to infer the biology of LUCA. For instance, if the ToL is governed by horizontal gene transfer (HGT), then it is not possible to infer the set of genes present in LUCA simply by looking at universally conserved genes (Doolittle 1999). This issue is because a widely distributed gene could have its origin in one cellular lineage and then be horizontally transferred to the other two lineages early in the evolution of life. One possibility to identify traits inherited from LUCA in the presence of HGT would be to look at those universally (or nearly universal) conserved genes having three domain phylogenies (i.e., phylogenies where Archaea, Bacteria, and Eukarya show as monophyletic). The first attempt in this direction resulted in 50 gene families, most of them related to the ribosome (Harris et al. 2003). The existence of ribosomes indicates that LUCA was capable of protein synthesis and therefore was more complex than a *progenote*.

3.2   The Nature of LUCA and Its Conditions of Existence

In relation to its optimal growth temperature, living organisms could be classified as psychrophiles (≤ 15 °C), mesophiles (~15–45 °C), thermophiles (~ 45–80 °C), and hyperthermophiles (≥ 80 °C) (Gaucher et al. 2010). The ToL, as revealed by rRNA, enabled the ancestral states of present-day characters to be inferred. By mapping the location of heat-loving prokaryotes along the ToL, Stetter proposed in 1996 that LUCA was a hyperthermophile (Stetter 1996). Accordingly, the deepest and short- est branches of the tree were occupied by prokaryotes living at >70 °C: *Aquifex* and *Thermotoga* (among the Bacteria) and *Pyrodictium, Pyrobaculum, Desulfurococcus, Sulfolobus, Methanopyrus, Thermococcus, Methanothermus*, and *Archaeoglobus* (within the Archaea).

However, at the end of the last century, an intense debate began regarding the temperature at which LUCA lived. At that time, the prevailing view was that LUCA was a hyperthermophile (Stetter 1996). Nevertheless, statistical analysis of rRNA suggested that LUCA was a mesophile or a thermophile at best (Galtier et al. 1999). This analysis was based in the correlation between G + C content of rRNA stems and known optimal growth temperature of prokaryotes (Galtier and Lobry 1997). The above analysis was contested by Di Giulio (2000) by using a series of rules based on parsimony to infer the G + C content of rRNA in LUCA. According to Di Giulio, the universal ancestor was a hyperthermophile. A few years later, the posi- tion of hyperthermophiles at the base of the ToL was questioned by Brochier and Philippe (2002). By removing fast-evolving unreliable sites from the multiple align- ment of rRNA sequences, Brochier and Philippe (2002) obtained a ToL in which *Aquifex* and *Thermotoga* were no longer the deepest branching taxa. Instead, the *Planctomycetales* occupied the basal position among Bacteria. This result suggested that hyperthermophilic bacteria adapted to high temperatures by receiving genes from heat-loving Archaea through HGT. However, the basal position of *Planctomycetales* was questioned also by Di Giulio (Di Giulio 2003; Barion et al. 2007), and, more recently, novel hyperthermophilic bacteria have been discovered branching at the base of the bacterial tree, earlier than the appearance of *Aquifex* and *Thermotoga* (Takami et al. 2012; Colman et al. 2016).

Later on, Boussau et al. (2008) implemented a similar approach to Galtier et al. (1999) but this time using 456 organisms sampled from Bacteria, Archaea, and Eukarya. In addition to the analysis of G + C content in rRNA stems, Boussau et al. (2008) also used the correlation between optimal growth temperature and the frequency of the amino acids: IVYWREL (Zeldovich et al. 2007) in 56 nearly universal proteins to infer the optimal growth temperature of LUCA. Both rRNA and amino acid ancestral reconstruction resulted in a non-hyperthermophilic LUCA. Their analysis also revealed that the last bacterial common ancestor (LBCA) and the last archaeal common ancestor (LACA) were both hyperthermophilic. This result was confirmed by Groussin et al. (2013), who analyzed the phylogenetic signal present in molecular sequences implementing improved ancestral reconstruction models for rRNA and amino acid sequences.

3.3 Reverse Gyrase: A Clue to Trace the Evolution of Heat-Loving Microbes

The discovery of the enzyme reverse gyrase in *Sulfolobus acidocaldarius*, which is active only at temperatures >55 °C, immediately raised the suspicion that this molecule is necessary to prevent denaturation by heat of genomic DNA with poor G + C content (Kikuchi and Asai 1984). It also suggested that other hyperthermophiles would require the enzyme for living at high temperatures. Several years after the discovery of reverse gyrase, Forterre (2002) took a genomic comparative approach to look for hyperthermophile-specific proteins. Surprisingly, he found that the only hyperthermophile-specific protein was reverse gyrase.

In principle, it would be straightforward to see if the phylogenetic distribution and evolution of reverse gyrase is consistent with its presence in the genome of LUCA. Recently, Weiss et al. (2016) attempted a reconstruction of the gene complement of LUCA by looking for those gene families that fulfill two criteria: (1) *the protein should be present in at least two higher taxa of Bacteria and Archaea*, and (2) *its tree should recover bacterial and archaeal monophyly*. The results revealed 355 gene families that putatively were present in LUCA, among which there was the reverse gyrase. Therefore, Weiss et al. (2016) suggested that LUCA was indeed a hyperthermophile. However, the selection criteria used by Weiss et al. (2016) were criticized by Gogarten and Deamer (2016) as perhaps being biased toward genes "with a limited distribution and utility in today's organisms" like those used by cells to live at higher temperatures. Gogarten and Deamer argued that genes used by most cells are more likely to be horizontally transferred between Archaea and Bacteria (Fig. 3.2a) and thus not selected by the methodology of Weiss et al. (2016), whereas genes used only by thermophiles are likely to be selected by the specified criteria if "these were transferred between the domains and, subsequently, between orders of

A)

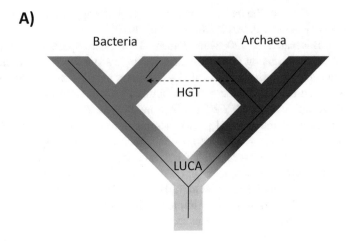

B)

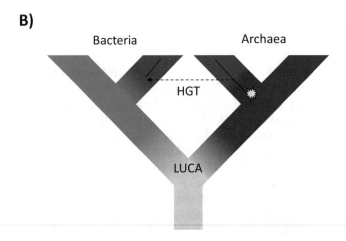

Fig. 3.2 False positives and negatives when inferring the genome of LUCA. (**a**) Highly conserved genes showing evidence of HGT are not identified to be inherited from LUCA and thus represent false-negative cases; (**b**) genes originated in one lineage (e.g., in Archaea: yellow star) and subsequently transferred to Bacteria are identified as inherited from LUCA and therefore represent false positives. Purple branches denote hyperthermophilic prokaryotes

thermophiles within the same domain" (Fig. 3.2b). The result, according to Gogarten and Deamer, was preliminary and highly biased towards a set of genes present in a minority of genomes in the tree of life.

This result was in close agreement with a recent result found by Catchpole and Forterre (2019) with respect to reverse gyrase evolution. In their analysis, which comprises the largest reverse gyrase dataset to date (376 sequences), archaeal and bacterial lineages do not form monophyletic groups, thereby contradicting the result presented by Weiss et al. (2016). Moreover, the short branch lengths separating archaeal from bacterial lineages in the reverse gyrase tree are not consistent with the long branches separating Archaea from Bacteria in phylogenies made from universal markers like 16S rRNA gene, RNA polymerases, or EF-G transcription factors, thus suggesting that reverse gyrase has not been inherited from the universal ancestor (Fig. 3.3). This analysis reinforces the idea that reverse gyrase evolved after the time of LUCA, perhaps in Archaea, and was later transferred to hyperthermophilic Bacteria.

3.4   Ancestral Protein Reconstruction: A Paleobiochemistry Experimental Approach

Ancestral sequence reconstruction enables the testing of hypotheses concerning the past histories of protein families. From multiple alignments of homologous sequences and tree topology, ancestral sequences can be inferred at each node in the tree by using a model of sequence evolution. Once inferred, ancestral sequences can be synthesized in vitro to study their properties. This approach has been used to reconstruct ancient proteins of earliest organisms and trace back the changes in temperature of the ancient biosphere (Gaucher et al. 2008; Akanuma 2017).

For example, the unfolding temperature of nucleoside diphosphate kinase (NDK) correlates with the optimal environmental temperature of its host organism (Akanuma et al. 2013). In an attempt to infer the temperature preferences of LUCA, Akanuma et al. (2013) reconstructed the ancestral protein sequence of bacterial and archaeal NDK with a homogeneous model of sequence evolution. They found that the ancestral bacterial and archaeal NDK sequences had a thermal stability equivalent to or greater than those of extant homologous from thermophilic organisms. Later on, Akanuma et al. (2015) repeated their analysis now with a nonhomogeneous model of sequence evolution and excluding hyperthermophilic sequences from their dataset. They found that the ancestral proteins from Bacteria and Archaea are still thermally stable.

In their analysis, Akanuma et al. (2013) were not able to directly reconstruct the NDK sequence of LUCA because of the lack of a root in the NDK tree. To infer the NDK sequence of LUCA, Akanuma and coworkers used the following heuristic approach (Akanuma et al. 2013). The ancestral archaeal and bacterial NDK differed by only 24 residues. They reasoned that the NDK of LUCA would contain some combination of these 24 residues. To test the hypothesis of a heat-loving LUCA, they first took the ancestral bacterial NDK (because it had the lowest unfolding temperature) and identified those substitutions among the 24 variants, which

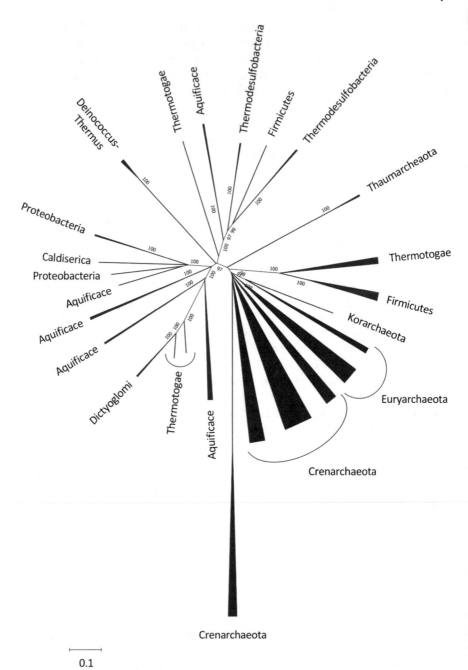

Fig. 3.3 Phylogenetic history of reverse gyrase. This enzyme is likely to have originated in the Archaea and vertically inherited in this domain. However, in Bacteria the phylogeny is more consistent with multiple HGT events. The short branch separating Archaea from Bacteria and the lack of monophyly of the domains do not suggest the presence of reverse gyrase in LUCA. Node numbers refer to 100 bootstrap replicas

decreased its thermal stability when introduced individually. By this they found seven destabilizing substitutions. Second, they introduced all seven substitutions simultaneously into the ancestral bacterial NDK and measured its thermal stability. They found that the unfolding temperature of this NDK was 94 °C, which corresponds to an environmental temperature of 75 °C. They suggested that this is the lowest estimate for the environmental temperature of LUCA.

Is the inference made from the NDK analysis reliable to infer the environmental temperature of LUCA? In the first place, we should ask whether NDK was indeed present in LUCA. Given the function of NDK, it is very likely to be a very ancient enzyme already present in LUCA. However, the phylogenetic analysis of NDK presented by Akanuma et al. (2013) shows the Archaea domain as polyphyletic. Although the polyphyly of Archaea in NDK phylogeny is likely the result of HGT after the split of Archaea from Bacteria, other positions of the root should be explored more thoughtfully. In any case, ancestral protein reconstructions of more enzymes are needed to evaluate if there is a general trend of increased thermal stability of the LUCA proteome.

3.5   When Did LUCA Live?

The solar system began to form ~4.56 billion years ago (Gya) and the Earth-Moon system originated when the proto-Earth collided against Theia ~4.53 Gya (Crawford 2013). During the following ~600 million years, the Earth experienced a series of events including the Late Heavy Bombardment. The formation of the Moon marks a maximum constraint for the age of life on Earth. The oldest, however disputed, evidences of ancient life include microfossils from the Nuvvuagittuq Belt in Quebec, Canada (3770–4280 Gya) (Dodd et al. 2017), cell-like inclusions (3800 Gya) (Pflug and Jaeschke-Boyer 1979), stromatolites (3700 Gya) (Nutman et al. 2016), and isotope data from the supracrustal belt in West Greenland (>3700 Gya) (Mojzsis et al. 1996; Rosing 1999) and a controversial evidence from preserved carbon in 4100 Gya zircons (Bell et al. 2015). However, the oldest indisputable evidence of life (according to Betts et al. (2018)) is the microfossils from the Strelley Pool Formation from Australia, having ~3.4 Gya (Sugitani et al. 2015).

Based on a recently calibrated two-domain phylogenetic tree (i.e., a tree where Eukarya branches from within the Archaea and the root are located between Archaea and Bacteria), Betts et al. (2018) suggest that LUCA lived >3.9 Gya; this is before the Late Heavy Bombardment ended. How does this time estimate accommodate with the inferences on the nature of LUCA and its predecessors? The impact-bottleneck hypothesis suggests that a large meteorite impact capable of heating the oceans by 100 °C eliminated all life on Earth except hyperthermophilic microorganisms capable of living at such high temperatures (Gogarten-Boekels et al. 1995). In this scenario, LUCA was not necessarily hyperthermophile, but the ancestor of the lineages that survived the hecatomb. Accordingly, the surviving lineages were hyperthermophilic Archaea and Bacteria (Nisbet and Sleep 2001). In fact, computer simulations of the Late Heavy Bombardment suggest the Earth was not completely sterilized during this time by meteoritic impacts (Abramov and Mojzsis 2009; Grimm and Marchi 2018). Additionally as discussed by Gogarten-Boekels (1995), the long branches connecting Archaea and Bacteria (i.e., the absence of deep branching lineages) can be explained by the catastrophic extinction made by a large meteorite impact.

Interestingly, the impact-bottleneck hypothesis fits the results provided by Boussau et al. (2008) and Groussin et al. (2013) regarding the prediction that the ancestor of Bacteria and Archaea was hyperthermophilic microbes while LUCA was a thermophile at best. It also fits the prediction made by the ancestral reconstruction of the NDK Akanuma et al. (2013). However, in this last case the ancestral reconstruction of the NDK suggests LUCA was a hyperthermophile.

Therefore, the timing provided by the molecular clock analysis is consistent with LUCA living before the Late Heavy Bombardment ended (Betts et al. 2018) (Fig. 3.4). With this result, it remains to be seen whether the position of hyperthermophiles in the bacterial tree is basal or not (Brochier and Philippe 2002; Barion et al. 2007) although the discovery of recent hyperthermophiles in the base of the bacterial tree suggests that this is the case (Takami et al. 2012; Colman et al. 2016).

3.6 Conclusions and Perspectives

Here, we show that the impact-bottleneck hypothesis accommodates various inferences regarding the growth temperature of LUCA and its immediate predecessors. This kind of argumentation used to reveal evolutionary history, where several facts are coordinated by a central explanation, is called *consilience* (or concordance of several) and was used by Darwin to suggest that all domestic pigeons originated from the wild *Columba livia* (Gould 2002). It can be said that the same kind of argumentation was used to infer the existence of the RNA world (Lazcano 2016). However, a close fit of the hypothesis to the data does not prove that the hypothesis is right. As suggested by the philosopher of biology Elliott Sober, if you hear a loud noise in the room above, you could suggest it was made by gremlins playing bowling in the attic; however, it is almost certainly not the case (Page and Holmes 1998). Therefore, in the case of the impact-bottleneck hypothesis discussed here, we can only expect that further data is still consistent with it. For instance, if hyperthermophiles don't lie at the base of the bacterial or archaeal branch (Brochier and Philippe 2002; Raymann et al. 2015), then the meteorite-bottleneck hypothesis loses explanatory power and is more likely to be wrong. Only further data will tell.

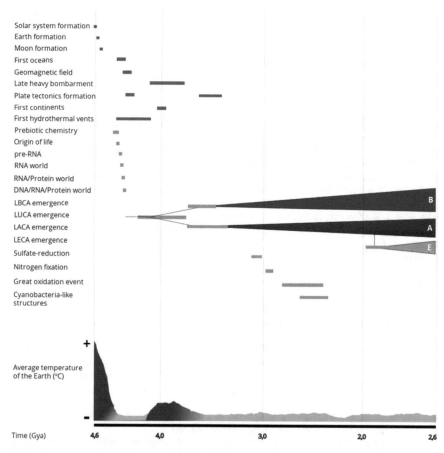

Fig. 3.4 Timeline of the early evolution of life on Earth. Bars represent a time estimated of every process. Data taken from the following sources: solar system formation ~4.6 Gya (Banerjee et al. 2016); Earth formation ~4.56 Gya (Maruyama and Ebisuzaki 2017); Moon formation ~4.53 Gya (Crawford 2013); first oceans ~4.3 Gya (Lyons et al. 2015); a Hadean geomagnetic field ~4.2 Gya (Tarduno et al. 2015); duration of the Late Heavy Bombardment (from ~4.1 to ~3.8 Gya) (Lowe and Byerly 2018); plate tectonics initiation ~ 4.4 Gya (Wilde et al. 2001); first continents formed at ~4 Gya (Hastie et al. 2016); oldest evidence of hydrothermal vent precipitates ~3,7–4,2 Gya (Dodd et al. 2017); timeframe for the origin and early diversification of life of ~300 million years (Lazcano and Miller 1996); emergence of LUCA ~3.9 Gya, LBCA ~3.4 Gya, LACA ~3.4 Gya, and LECA ~1.8 Gya (Betts et al. 2018); sulfate reduction (~3.0 Gya) and methanogenesis (~2.6 Gya) initial activity (Lyons et al. 2015); great oxygen event from ~3.8 to ~2.3 Gya (Lyons et al. 2014); *Cyanobacteria*-like structures at ~3.4 Gya (Schopf 1993). Archaeal (A), Bacterial (B), and Eukarya (E) domains were evolved through time. The average temperature curve of the Earth (at the bottom) is only for illustrative purposes and shows the increment of the temperature during the impact-bottleneck process

Acknowledgments The authors would like to thank María González for critical reading of the manuscript.

References

Abramov O, Mojzsis SJ (2009) Microbial habitability of the Hadean Earth during the late heavy bombardment. Nature 459(7245):419

Akanuma S (2017) Characterization of reconstructed ancestral proteins suggests a change in temperature of the ancient biosphere. Life 7(3):33

Akanuma S, Nakajima Y, Yokobori SI, Kimura M, Nemoto N, Mase T, Miyazono KI, Tanokura M, Yamagishi A (2013) Experimental evidence for the thermophilicity of ancestral life. Proc Natl Acad Sci 110(27):11067–11072

Akanuma S, Yokobori SI, Nakajima Y, Bessho M, Yamagishi A (2015) Robustness of predictions of extremely thermally stable proteins in ancient organisms. Evolution 69(11):2954–2962

Banerjee P, Qian YZ, Heger A, Haxton WC (2016) Evidence from stable isotopes and 10 Be for solar system formation triggered by a low-mass supernova. Nat Commun 7:13639

Barion S, Franchi M, Gallori E, Di Giulio M (2007) The first lines of divergence in the bacteria domain were the hyperthermophilic organisms, the Thermotogales and the Aquificales, and not the mesophilic Planctomycetales. Biosystems 87(1):13–19

Becerra A, Delaye L, Islas S, Lazcano A (2007) The very early stages of biological evolution and the nature of the last common ancestor of the three major cell domains. Annu Rev Ecol Evol Syst 38:361–379

Bell EA, Boehnke P, Harrison TM, Mao WL (2015) Potentially biogenic carbon preserved in a 4.1 billion-year-old zircon. Proc Natl Acad Sci 112(47):14518–14521

Betts HC, Puttick MN, Clark JW, Williams TA, Donoghue PC, Pisani D (2018) Integrated genomic and fossil evidence illuminates life's early evolution and eukaryote origin. Nat Ecol Evol 2(10):1556

Boussau B, Blanquart S, Necsulea A, Lartillot N, Gouy M (2008) Parallel adaptations to high temperatures in the Archaean eon. Nature 456(7224):942

Boyer M, Madoui MA, Gimenez G, La Scola B, Raoult D (2010) Phylogenetic and phyletic studies of informational genes in genomes highlight existence of a 4th domain of life including giant viruses. PLoS One 5(12):e15530

Brochier C, Philippe H (2002) Phylogeny: a non-hyperthermophilic ancestor for bacteria. Nature 417(6886):244

Catchpole R, Forterre P (2019) Positively twisted: the complex evolutionary history of reverse gyrase suggests a non-hyperthermophilic last universal common ancestor. bioRxiv:524215

Colman DR, Jay ZJ, Inskeep WP, Jennings RD, Maas KR, Rusch DB, Takacs-Vesbach CD (2016) Novel, deep-branching heterotrophic bacterial populations recovered from thermal spring metagenomes. Front Microbiol 7:304

Crawford I (2013) The Moon and the early Earth. Astron Geophys 54:1.31–1.34

Di Giulio M (2000) The universal ancestor lived in a thermophilic or hyperthermophilic environment. J Theor Biol 203(3):203–213

Di Giulio M (2003) The ancestor of the Bacteria domain was a hyperthermophile. J Theor Biol 224(3):277–283

Dodd MS, Papineau D, Grenne T, Slack JF, Rittner M, Pirajno F, O'Neil J, Little CT (2017) Evidence for early life in Earth's oldest hydrothermal vent precipitates. Nature 543(7643):60

Doolittle WF (1999) Phylogenetic classification and the universal tree. Science 284(5423):2124–2128

Forterre P (2002) A hot story from comparative genomics: reverse gyrase is the only hyperthermophile-specific protein. Trends Genet 18(5):236–237

Forterre P (2015) The universal tree of life: an update. Front Microbiol 6:717

Fournier GP, Huang J, Peter Gogarten J (2009) Horizontal gene transfer from extinct and extant lineages: biological innovation and the coral of life. Philos Trans R Soc B Biol Sci 364(1527):2229–2239

Freeland SJ, Knight RD, Landweber LF (1999) Do proteins predate DNA? Science 286(5440):690–692

Galtier N, Lobry JR (1997) Relationships between genomic G+C content, RNA secondary structures, and optimal growth temperature in prokaryotes. J Mol Evol 44(6):632–636

Galtier N, Tourasse N, Gouy M (1999) A non-hyperthermophilic common ancestor to extant life forms. Science 283(5399):220–221

Gaucher EA, Govindarajan S, Ganesh OK (2008) Palaeotemperature trend for Precambrian life inferred from resurrected proteins. Nature 451(7179):704

Gaucher EA, Kratzer JT, Randall RN (2010) Deep phylogeny--how a tree can help characterize early life on Earth. Cold Spring Harb Perspect Biol 2(1):a002238

Gogarten JP, Deamer D (2016) Is LUCA a thermophilic progenote? Nat Microbiol 1:16229

Gogarten-Boekels M, Hilario E, Gogarten JP (1995) The effects of heavy meteorite bombardment on the early evolution – the emergence of the three domains of life. Orig Life Evol Biosph 25(1–3):251–264

Gould SJ (2002) The structure of evolutionary theory. Harvard University Press, Cambridge

Grimm RE, Marchi S (2018) Direct thermal effects of the Hadean bombardment did not limit early subsurface habitability. Earth Planet Sci Lett 485:1–8

Groussin M, Boussau B, Charles S, Blanquart S, Gouy M (2013) The molecular signal for the adaptation to cold temperature during early life on Earth. Biol Lett 9(5):20130608

Harris JK, Kelley ST, Spiegelman GB, Pace NR (2003) The genetic core of the universal ancestor. Genome Res 13(3):407–412

Hastie AR, Fitton JG, Bromiley GD, Butler IB, Odling NW (2016) The origin of Earth's first continents and the onset of plate tectonics. Geology 44(10):855–858

Kikuchi A, Asai K (1984) Reverse gyrase – a topoisomerase which introduces positive superhelical turns into DNA. Nature 309(5970):677

Koonin EV, Wolf YI, Puigbò P (2009) The phylogenetic forest and the quest for the elusive tree of life. In: Cold Spring Harbor symposia on quantitative biology, vol 74. Harbor Laboratory Press, Cold Spring, pp 205–213

Lake JA, Henderson E, Oakes M, Clark MW (1984) Eocytes: a new ribosome structure indicates a kingdom with a close relationship to eukaryotes. Proc Natl Acad Sci U S A 81(12):3786–3790

Lazcano A (2016) The RNA world: piecing together the historical development of a hypothesis. Mètode Sci Stud J Ann Rev 6:166–173

Lazcano A, Miller SL (1996) The origin and early evolution of life: prebiotic chemistry, the pre-RNA world, and time. Cell 85(6):793–798

Lowe DR, Byerly GR (2018) The terrestrial record of late heavy bombardment. New Astron Rev 81:39–61

Lyons TW, Reinhard CT, Planavsky NJ (2014) The rise of oxygen in Earth's early Ocean and atmosphere. Nature 506(7488):307–315

Lyons TW, Fike DA, Zerkle A (2015) Emerging biogeochemical views of Earth's ancient microbial worlds. Elements 11(6):415–421

Maruyama S, Ebisuzaki T (2017) Origin of the Earth: a proposal of new model called ABEL. Geosci Front 8(2):253–274

Mojzsis SJ, Arrhenius G, McKeegan KD, Harrison TM, Nutman AP, Friend CR (1996) Evidence for life on Earth before 3,800 million years ago. Nature 384(6604):55

Moreira D, López-García P (2015) Evolution of viruses and cells: do we need a fourth domain of life to explain the origin of eukaryotes? Philos Trans R Soc B Biol Sci 370(1678):20140327

Nisbet EG, Sleep NH (2001) The habitat and nature of early life. Nature 409(6823):1083

Nutman AP, Bennett VC, Friend CR, Van Kranendonk MJ, Chivas AR (2016) Rapid emergence of life shown by discovery of 3,700-million-year-old microbial structures. Nature 537(7621):535

Page RDM, Holmes EC (1998) Molecular evolution: a phylogenetic approach. Wiley-Blackwell, Oxford/Malden

Patel BH, Percivalle C, Ritson DJ, Duffy CD, Sutherland JD (2015) Common origins of RNA, protein and lipid precursors in a cyanosulfidic protometabolism. Nat Chem 7(4):301

Pflug HD, Jaeschke-Boyer H (1979) Combined structural and chemical analysis of 3,800-Myr-old microfossils. Nature 280(5722):483

Raymann K, Brochier-Armanet C, Gribaldo S (2015) The two-domain tree of life is linked to a new root for the Archaea. Proc Natl Acad Sci U S A 112(21):6670–6675

Rivera MC, Lake JA (2004) The ring of life provides evidence for a genome fusion origin of eukaryotes. Nature 431(7005):152

Rosing MT (1999) ^{13}C-depleted carbon microparticles in 3700-Ma Sea-floor sedimentary rocks from West Greenland. Science 283:674–676

Sagan L (1967) On the origin of mitosing cells. J Theor Biol 14(3):225–IN6

Schopf JW (1993) Microfossils of the early Archean apex chert: new evidence of the antiquity of life. Science 260(5108):640–646

Spang A, Saw JH, Jørgensen SL, Zaremba-Niedzwiedzka K, Martijn J, Lind AE, van Eijk R, Schleper C, Guy L, Ettema TJ (2015) Complex archaea that bridge the gap between prokaryotes and eukaryotes. Nature 521(7551):173

Stetter KO (1996) Hyperthermophilic procaryotes. FEMS Microbiol Rev 18(2–3):149–158

Sugitani K, Mimura K, Takeuchi M, Lepot K, Ito S, Javaux EJ (2015) Early evolution of large micro-organisms with cytological complexity revealed by microanalyses of 3.4 Ga organic-walled microfossils. Geobiology 13(6):507–521

Takami H, Noguchi H, Takaki Y, Uchiyama I, Toyoda A, Nishi S, Chee GJ, Arai W, Nunoura T, Itoh T, Hattori M (2012) A deeply branching thermophilic bacterium with an ancient acetyl-CoA pathway dominates a subsurface ecosystem. PLoS One 7(1):e30559

Tarduno JA, Cottrell RD, Davis WJ, Nimmo F, Bono RK (2015) A Hadean to Paleoarchean geodynamo recorded by single zircon crystals. Science 349(6247):521–524

Weiss MC, Sousa FL, Mrnjavac N, Neukirchen S, Roettger M, Nelson-Sathi S, Martin WF (2016) The physiology and habitat of the last universal common ancestor. Nat Microbiol 1(9):16116

Wilde SA, Valley JW, Peck WH, Graham CM (2001) Evidence from detrital zircons for the existence of continental crust and oceans on the Earth 4.4 Gyr ago. Nature 409(6817):175

Woese CR, Fox GE (1977a) Phylogenetic structure of the prokaryotic domain: the primary kingdoms. Proc Natl Acad Sci U S A 74:5088–5090

Woese CR, Fox GE (1977b) The concept of cellular evolution. J Mol Evol 10(1):1–6

Woese CR, Kandler O, Wheelis ML (1990) Towards a natural system of organisms: proposal for the domains Archaea, bacteria, and Eucarya. Proc Natl Acad Sci U S A 87(12):4576–4579

Zeldovich KB, Berezovsky IN, Shakhnovich EI (2007) Protein and DNA sequence determinants of thermophilic adaptation. PLoS Comput Biol 3(1):e5

Chapter 4
Stromatolites, Biosignatures, and Astrobiological Implications

Jamie S. Foster, Joany Babilonia, Erica Parke-Suosaari, and R. Pamela Reid

Abstract For millennia, humanity has looked to stars and wondered, "Are we alone in the universe?" Although this question was initially the purview of philosophers, now, with leaps in scientific and technological advances, we have changed the nature of this question from existential to empirical. Today, the question "Are we alone?" serves as a crux to the field of astrobiology. To search for life elsewhere in the universe, we must first understand how life originates and evolves on Earth but also how biology leaves behind residual signatures of its existence. To address these questions, many astrobiology researchers have targeted stromatolite-forming communities as model ecosystems to explore how microbe–mineral interactions, under a range of environmental conditions, can lead to the formation of biosignatures. Stromatolites are depositional structures formed by the activities and interactions of microbes and have a fossil record dating back billions of years. Due to their long evolutionary history and abundance on the modern Earth, research on the biological, chemical and geological processes of stromatolite formation have provided important insights into the field of astrobiology, including the diversity and preservation of biosignatures. In this chapter, we examine the range of biosignatures found in stromatolites and how these markers improve our understanding of the past, present, and future of life in the context of astrobiology. We also discuss whether stro-

J. S. Foster (✉) · J. Babilonia
Department of Microbiology and Cell Science, Space Life Sciences Lab, University of Florida, Merritt Island, FL, USA
e-mail: jfoster@ufl.edu

E. Parke-Suosaari
Rosenstiel School of Marine and Atmospheric Science, University of Miami, Miami, FL, USA

Department of Mineral Sciences, National Museum of Natural History, Smithsonian Institution, Washington, DC, USA

Bush Heritage Australia, Melbourne, VIC, Australia

R. P. Reid
Rosenstiel School of Marine and Atmospheric Science, University of Miami, Miami, FL, USA

matolite research can play a role in the future exploration of habitable worlds in our own solar system and beyond.

4.1 Stromatolites Are Ideal Models for Astrobiology Research

Stromatolites are controversial. Scientists argue about their age, their origin, and whether modern systems can serve as adequate analogs to understanding the earliest habitats on Earth. Despite the academic discord, both fossil and living stromatolites represent valuable resources to understand not only Earth's past but also current feedbacks that occur between life and its environment. Stromatolites are a type of microbialite and are one of the most prevalent and most recognized constituents of the Precambrian carbonate platforms (Grotzinger and Knoll 1999; Reid et al. 2011; Suosaari et al. 2016b, 2019). Stromatolites are laminated, lithified organosedimentary structures that have formed as a result of the sediment trapping, binding, and precipitating activities of microbes in response to their local environment (Awramik et al. 1976). As products of interactions between microbes, minerals and the environment, stromatolites are a quintessential embodiment of linkages between the biosphere and the geosphere. The interplay between biological and geological activities resulting in stromatolite formation has had a profound influence on the evolution and function of life on this planet.

For billions of years, stromatolite-forming microbial communities have played a major role in regulating global cycles of major elements and sedimentation via production and decomposition of organic matter, trapping and binding of sediment, and precipitation of calcium carbonate. As such, stromatolites are ideal model systems in which to investigate key astrobiological questions regarding the interactions between microbial life and the environment as well as how these interactions shape habitability.

4.2 Long Live the Stromatolites and Their Biosignatures

Stromatolites have an extensive geological record extending deep into the Archean. Recent discoveries have pushed the boundaries of well-organized microbial ecosystems back as far as 3.7 Ga, with reports of stromatolite-like ecosystems within the Isua supracrustal belt located in southwest Greenland (Nutman et al. 2019). Although the biogenicity of this site has been challenged (Allwood et al. 2018), newly published geological and geochemical evidence reinforces the idea that the Isua structures are likely biogenic in origin (Nutman et al. 2019). The fossil record indicates that by the late Archean, stromatolites became increasingly common, forming expansive carbonate reef systems reaching a peak in abundance and diversity during the Proterozoic (Awramik 1991; Reid 2011). For example, well-characterized stromatolite-rich carbonates have been found at the boundary of the late Archean and Proterozoic eons (2.5 Ga) in the Campbell and Subgroup of South Africa (Sumner and Bowring 1996; Kamber and Webb 2001) as well as within the laminated 3.4 Ga formations of the eastern Pilbara Craton of Western Australia (Lowe 1980; Walter et al. 1980; Van Kranendonk et al. 2003), thus providing important insights into dynamic changes of Earth's environment at key evolutionary time points.

Due to the long evolutionary history of stromatolites, these microbial ecosystems serve as important reservoirs for biosignatures. Biosignatures are patterns or products made by the activity of life and not by abiotic processes (for review see Hays et al. 2017; Chan et al. 2019). As the field of astrobiology is hindered by the fact that we only have one representative biosphere to develop tools and hypotheses about the nature of life in the universe, cognitive biases that impact our ability to recognize and detect extraterrestrial life can be enormous. With those biases in mind, it is likely that all life generates some type of residual biosignature, and stromatolites can provide a valuable ecosystem to explore the processes by which biosignatures form under a dynamically changing environment (Fig. 4.1). Here, we briefly discuss the patterns (e.g., morphology and fabrics) and products (e.g., minerals and molecules) of stromatolites as they pertain to the study of biosignatures across different spatial scales, as well as the role that stromatolite research can play in searching for the constraints of habitability beyond Earth.

4.3 Stromatolite Patterns as Biosignatures: Morphology and Fabrics

Morphological analysis of stromatolites can provide important insight into the feedbacks, or cause-and-effect relationships, between biotic and environmental evolution. Spatial self-organization has important implications regarding ecosystem function, specifically regarding resilience and productivity (Rietkerk and van de Koppel 2008). By organizing themselves into regular spatial patterns, ecological communities can concentrate local resources to create a more robust ecosystem that can resist environmental perturbations.

Stromatolites are ideal ecosystems to explore the feedbacks between life and the environment and how these feedbacks can lead to morphological biosignatures. The dominance of stromatolites in the geologic record, and their persistence over time, suggests that the stromatolite-building microbial communities have a selective advantage that benefits community structure to withstand disturbances. As stromatolites grow, they generate geomorphological features capable of influencing biological processes, which in turn exacerbate physical processes and induce net negative feedbacks (depending on the density of the organisms), leading to regular morphological patterns (Rietkerk and van de Koppel 2008; Budd et al. 2016). These

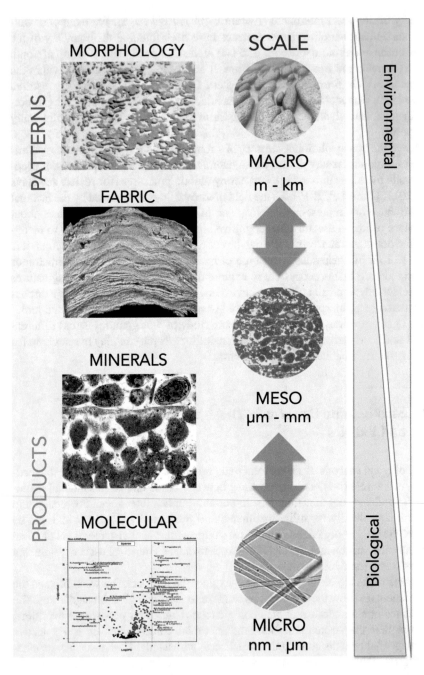

Fig. 4.1 Stromatolites biosignatures occur across multiple scales and serve as valuable resources for astrobiology research. Stromatolites biosignatures can be loosely categorized as "patterns and products." Characteristics, such as overall morphology and fabric structures, can provide insight into the environmental inputs, whereas micro- and mesoscale products of stromatolite growth, such as precipitated carbonate and metabolic outputs, can elucidate the underlying biological processes of biosignature, and ultimately stromatolite formation. Together, these patterns and products can provide important insight into the formative environmental and biological influences across multiple spatial and temporal scales

self-organizing patterns may be indicative of life processes, thereby serving as important biosignatures.

Spatial patterns in the environment may reflect feedback loops that are dependent on scale (i.e., short or long distances) (Rietkerk and van de Koppel 2008) and homogenous conditions that lack strong environmental gradients (Turing 1952). For example, Turing's (Turing) activator–inhibitor principle suggests that resulting patterns are not imposed on the system, but instead emerge as a result of various feedbacks. Positive feedbacks, where organisms help each other survive through changing the local environment, dominate the ecosystem at short distances; and therefore, aggregation of microbes may provide key functions for the community, such as facilitating nutrient retention and helping to reduce community losses to predation or wave action (Liu et al. 2012). Negative feedbacks, where organisms deplete resources and are in competition with each other, dominate at larger distances (Rietkerk and van de Koppel 2008). Alternatively, when the environment is heterogeneous or heavily grazed, structures may lose self-organizing patterns or pattern formation can be obscured (Rietkerk et al. 2004; Rietkerk and van de Koppel 2008). Short-distance positive feedbacks may increase the sharpness of organizational patterns, but without the coupling of long-distance negative feedback, formation of regular patterns may not be possible (van de Koppel and Crain 2006).

An ideal ecosystem to explore the role of biogenic spatial patterns as biosignatures has been the stromatolites of Shark Bay in Western Australia (Suosaari et al. 2016a, 2019). The first documented examples of modern microbial build-ups with sizes and shapes comparable to Precambrian forms were discovered in 1954 in a hypersaline embayment of Shark Bay, Western Australia, known as Hamelin Pool (Playford et al. 2013). With a shoreline of 135 km made up almost entirely of microbial mats and stromatolites, Hamelin Pool is an iconic location hosting the world's most extensive and diverse modern marine stromatolite system.

A recent field-intensive mapping effort was conducted to examine stromatolite distribution within the site (Suosaari et al. 2016a). This mapping program differed from previous studies (Jahnert and Collins 2012) in that stromatolites were classified on the basis of morphology rather than microbial mat type or surface texture (Fig. 4.2). Additionally, high-resolution single-beam sonar data were combined with Landsat 8 satellite imagery to produce a detailed bathymetry map, which revealed a unique geographic distribution of morphologically distinct stromatolite structures around the margins of the Pool (Suosaari et al. 2019). This geographic zonation allowed the differentiation of eight "Stromatolite Provinces," each with distinct patterns of stromatolite morphology and unique shelf physiography (Suosaari et al. 2016b, 2019).

The microbial surface mats that generate the underlying stromatolite structures in Hamelin Pool produce exopolymeric substances (EPS), which enhance sediment cohesiveness and stability (Blanchard et al. 2000), and in turn help structures accumulate and grow above wave-base. When elevated above sediment abrasion, and with enough accommodation space, the surface area increases and adjacent structures may merge together. Stromatolites can also influence currents, with decreased water depth increasing water flow by channeling the flow between structures.

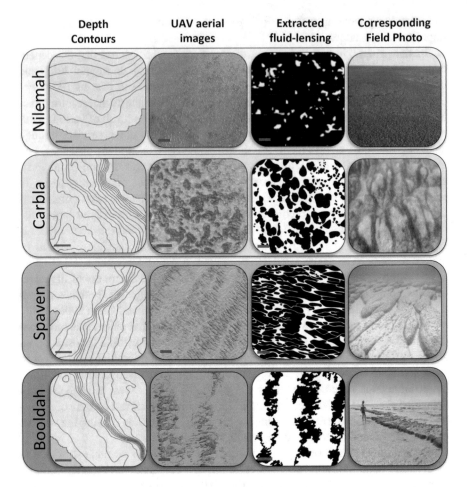

Fig. 4.2 Morphological self-organizing patterns may be indicative of life processes, thereby serving as important biosignatures. The different Provinces of Hamelin Pool exhibit highly variable morphologies that are dependent on location with the pool. The first column displays depth contours from a selected area within the Provinces of Nilemah, Carbla, Spaven, and Booldah (Bar, 1 km). The second column shows an example of sub-centimeter scale 2D imagery collected via UAV platform and processed using Fluid Lensing (Bar, 2 m). Column three displays highly distinctive morphological structures extracted structures from the fluid-lensed imagery (Bar, 1 m). Column four provides representative field photos of the targeted Provinces

Exposed stromatolite surfaces can also stabilize sediment, promoting microbial growth, further increasing ecosystem stability. Runnels, or drainage channels, focus the increased water flow, preventing colonization of additional microbial communities. Observations of stromatolite morphologies (Fig. 4.2; Suosaari et al. 2016a, 2019) suggest that these lower level feedbacks may lead to distinctive self-organizing patterns (Camazine et al. 2003; Budd et al. 2016) and are highly influenced by the energy regimes of the local environment (Suosaari et al. 2016a, 2019).

Whereas the morphology and physiography of stromatolites provide important insight into the environmental conditions of the ecosystem, the internal fabrics of stromatolites can help delineate the biological influences on formation and serve as yet another distinctive biosignature (Logan et al. 1974; Reid et al. 2000, 2003). For decades, analyses of the microfabrics in stromatolites have been successfully used to interpret the nature and metabolisms of the microbial mat communities that produce them (Visscher et al. 1998; Reid et al. 2000, 2003; Pace et al. 2018). Mineral precipitation in lithifying microbial mat ecosystems is a complex, intricate process that is dependent on the saturation index as well as the presence of potential nucleation sites (e.g., EPS) within the mat community (Dupraz et al. 2009). Net precipitation occurs only if the processes of enhancing carbonate precipitation (i.e., microbial guilds that promote precipitation) outweigh the processes that enhance the dissolution of carbonate minerals (i.e., microbial guilds that promote dissolution). Additionally, features within the microfabrics, such as laminae, can form from the successive cycling of microbial surface communities resulting in depositional events of cemented and uncemented sediments coupled with the formation of micrite (Reid et al. 2000). The specific characteristics of the microfabrics, whether they were the product of trapping and binding or micrite precipitation, are highly informative of the type of microbial community that made the structure. For example, in Bahamian stromatolites, fused micritized sand grains are the product of endolithic coccoid cyanobacteria. Endoliths can cross through organic matrix material (e.g., EPS) and bore into the sand grains, which becomes infilled with micrite, essentially fusing the grains and forming cemented layers (Reid et al. 2011). The nature and characteristics of the microfabrics can then be used as an important biosignature to infer the biological processes associated with its formation over time thereby improving the interpretation of the fossil record (Frantz et al. 2015; Pace et al. 2018).

4.4 Stromatolite Products as Biosignatures: Minerals

Of the more than 5300 mineral species found on Earth, the vast majority are the products of interactions and alterations by life, with approximately only 1500 minerals, the result of purely abiotic processes (Chan et al. 2019). However, with only a single biosphere as a reference point, one cannot assume that the processes by which these biogenic minerals form on Earth are the same on other planets and moons. For example on Earth, the reservoirs of hydrocarbons, or fossil fuels, are typically derived from geologically altered biomass and found in Earth's interior, whereas on Titan, the lacustrine hydrocarbon deposits are located on the surface and are thought to be abiotic in origin, a product of photochemical interactions with Titan's nitrogen-rich atmosphere (Lorenz et al. 2008; Cornet et al. 2015). Therefore, the presence of a given mineral cannot be considered solely as evidence for life; however, it can inform the environmental context and conditions in which the mineral formed and may serve as an indirect biosignature of life.

Many of the deposited minerals found in stromatolites, such as carbonates, halite, and gypsum, are found on other terrestrial worlds, such as Mars (Aubrey et al. 2006; Masse et al. 2012; Niles et al. 2013) and do not directly indicate life. However, chemical alterations of minerals, such as isotope fractionation, can be used in conjunction with the depositional setting to help characterize whether the mineral is biotic in origin and may provide insight into the metabolism that facilitated the process. Isotopic fractionation in stromatolites has long been used to understand the biological processes associated with its deposition; however, the fractionation profiles can vary considerably depending on the isotope examined and the environmental context (Andres and Reid 2006; Breitbart et al. 2009; Planavsky et al. 2009; Chagas et al. 2016; Louyakis et al. 2017). For example, stable sulfur isotopes have been used to understand and reconstruct stromatolite formation in both ancient and modern systems due to pronounced fractionation patterns caused by sulfate-reducing microbes (Kilburn and Wacey 2011; Bontognali et al. 2012). Sulfate-reducing microbes preferentially use ^{32}S relative to ^{34}S, thereby releasing sulfides that are depleted in the heavier isotope. Microbes can also deplete sulfur of various oxidation states, which can also impact the isotopic fractionation values (Bontognali et al. 2012; Weber et al. 2016). Microbial-induced transformations of sulfur are ancient (Shen et al. 2001) and sulfur-metabolizing microbes have been linked to stromatolite formations that date back 3.45 Ga (Bontognali et al. 2012). In modern stromatolites, the formation of the laminae characteristic of stromatolites is also highly correlated with sulfur cycling. For example, in the Bahamian stromatolites of Highborne Cay and Little Darby Island, heterotrophic processes, such as sulfate reduction and sulfide oxidation coupled with photosynthesis, are major drivers of carbonate precipitation (Visscher et al. 1998; Reid et al. 2000; Andres and Reid 2006; Casaburi et al. 2016).

Carbon isotopic shifts can also provide important insight into how microbial communities cycle CO_2 and influence the $\delta^{13}C$ signatures, thereby recording past metabolic activities and influences on the environment (Andres et al. 2006). Stromatolites are ideal models to examine how shifts in $\delta^{13}C$ become recorded into the rock record under different environmental conditions, such as lacustrine or hypersaline. A comparison of the open marine stromatolites and thrombolites of Highborne Cay, Bahamas, which are only a few meters away from each other, revealed pronounced differences in $\delta^{13}C$ values (Andres et al. 2006; Planavsky et al. 2009; Louyakis et al. 2017). The intertidal thrombolites exhibited much heavier values compared to the adjacent stromatolites and likely reflect the larger role of heterotrophic processes in the formation of the subtidal stromatolite laminae. Carbon and sulfur fractionation patterns, coupled with their geological context may be valuable indirect biosignatures for the search for life beyond Earth. Sulfur deposits and carbonates have been found on the surface of Mars and derived meteorites (Aubrey et al. 2006). Additionally, close spectral matches of sulfate minerals have been observed on more distant moons, such as Europa (McCord et al. 1999; Orlando et al. 2005), indicating the importance and the need to more fully understand microbe–mineral interactions on Earth to help delineate what constitutes a biosignature under a given environmental context (Chan et al. 2019).

Another mineral found in stromatolites that has the potential to serve as a valuable biosignature is magnetite. A hallmark of modern living stromatolites is the trapping and binding activity of surface microbial mat communities (Reid et al. 2000). The abundance of EPS material on the mat surface can sequester sediment, which effectively helps stabilize the stromatolite-forming communities in high-energy wave environments and under steep angles (Braissant et al. 2007; Dupraz et al. 2009; Bowlin et al. 2012; Flood et al. 2014). In sedimentary environments, including carbonate-depositing stromatolite systems, fine detrital magnetite can be found, which is susceptible to external magnetic fields (Lund et al. 2010; Petryshyn et al. 2016). As stromatolites accrete, the trapping, binding, and precipitation activities can cause distinctive patterns in the magnetite abundance, which might not otherwise occur in the absence of a microbial mat community (Petryshyn et al. 2016). Research on microbialites from Tahiti (Lund et al. 2010) revealed a distinctive pattern of detrital magnetite grains bound within the framework of the microbialite and not in the surrounding abiotic carbonate grains, suggesting that the microbialite-forming communities, even at steep angles, were able to effectively generate a distinctive mineral biosignature pattern (Petryshyn et al. 2016).

4.5 Stromatolite Products as Biosignatures: Organic Molecules

In addition to the type of mineral present, minerals within stromatolites can also serve as hubs or attachment points for molecular biosignatures (Chaçon 2010; Alleon and Summons 2019). Minerals have long been thought to serve as hot spots for the origin of life (Hazen and Sverjensky 2010), as the crystalline surfaces of minerals can serve to concentrate, protect, and catalyze certain biological reactions (Roling et al. 2015). Stromatolites represent an ideal ecosystem to examine how microbe–mineral interactions can lead to the preservation of a diverse range of organic molecular biosignatures. Molecular biosignatures can been considered those organic compounds that have retained their original chemical structure providing information about the organisms that made the molecule and environmental factors that may have led to its synthesis (Ourisson et al. 1982; Summons et al. 2008). There has been extensive research on using modern stromatolite-forming mat communities to identify various organic molecular biosignatures, in particular with lipids (Summons et al. 1999, 2013; Jahnke et al. 2004, 2008; O'Reilly et al. 2017); however, other molecules could serve as viable targets including nucleic acids as well as primary and secondary metabolites.

The search for suitable candidates for molecular biosignatures can be exceedingly difficult as the potential for biologically derived molecules to be preserved over geological time can vary between environmental conditions and the chemical structure (Summons et al. 2008). Moreover, the reconstruction of the community dynamics that led to the synthesis and deposition of the targeted molecule may be difficult to trace over evolutionary time. The billions of years of evolution and the reshaping of genomes through horizontal gene transfer, primarily at the hand of viruses, has recast the "Tree of Life" into a far more tangled "Web of Life" (Olendzenski and Gogarten 2009; Koonin and Wolf 2012). Therefore, the search for molecular biosignatures needs to expand to include not only individual molecules that identify a given organism or targeted metabolism, but also ecosystem-level suites of metabolic products that can provide community-level information about the biological processes or conditions that formed the molecular pattern (Noffke et al. 2013).

Despite these challenges, several advancements in multi-omics techniques have been used to survey stromatolite-forming mats at the ecosystem level to identify organic molecules and their distributions within the community as potential biosignatures. The semi-closed nature of stromatolites coupled with their high levels of productivity and rapid cycling of major elements renders stromatolites amenable to meta-omics and systems biology approaches to identify potential organic molecular biosignatures. Stromatolites have been subject of numerous multiomic analyses to more fully characterize the genomes, transcriptomes, and metabolomes of stromatolite-forming taxa in a diverse range of habitats (Desnues et al. 2008; Breitbart et al. 2009; Khodadad and Foster 2012; Mobberley et al. 2012, 2015; Peimbert et al. 2012; Edgcomb et al. 2013, 2014; Saghaï et al. 2015; White et al. 2015, 2016, 2018; Cerqueda-Garcia and Falcon 2016; Ruvindy et al. 2016; Warden et al. 2016; Casaburi et al. 2016; Louyakis et al. 2017, 2018; Babilonia et al. 2018; Gomez-Acata et al. 2019). For example, a comparative metagenomic analysis of the stromatolites of Hamelin Pool revealed distinctive genetic enrichments within the stromatolite-forming mats that strongly correlated with water depth within the pool (Babilonia et al. 2018). For the intertidal stromatolite communities, there was an enrichment in photosynthesis-related genes, whereas the subtidal mats showed enrichments in more heterotrophic pathways, such as genes associated with sulfate reduction (Fig. 4.3). These gene patterns, and their products, create distinctive biomarkers, or indicators of the biological state or condition in which they form, to elucidate the potential adaptive strategies and synergistic interactions between stromatolite-forming communities and their environment. Whether these distinctive genetic or metabolic biomarkers within different populations of stromatolites can be preserved and used as biosignatures over time remains to be determined. Regardless, the plethora of deciphered genes, metabolites and pathways are helping to lay an important foundation to understand the biological mechanisms associated stromatolite formation and how these cellular processes are impacted by perturbations to the environment.

Together, these recent advances in high-throughput microbiome and exposome analysis techniques coupled with the breakthrough field of paleogenomics represent a dynamic frontier for future molecular biosignature research. Although the application of paleogenomics, or the study of ancient nucleic acids, to microbial ecology is

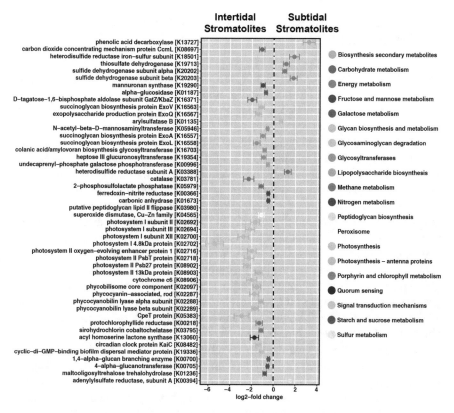

Fig. 4.3 Comparative metagenomics can create distinctive patterns of gene enrichments that provide insight into the environmental conditions that drive biological patterns. A direct examination of the metagenomes derived from an intertidal and subtidal stromatolite-forming microbial mat in Hamelin Pool, Shark Bay, Western Australia. Colors reflect clusters of pathways group by general function (i.e., KEGG pathways). A positive $\log_2$-fold change designates significantly gene abundances in the subtidal stromatolites, whereas a negative $\log_2$-fold change designates differential abundance within intertidal stromatolite-forming mats. These patterns can provide insights into potential organic molecular biosignature targets at the community or ecosystem level (Modified from Babilonia et al. 2018)

still in the early stages, with most research focusing on host–microbe interactions (Shapiro and Hofreiter 2014; Gorgé et al. 2016), efficient technical and computational screening capabilities to examine long-lived residual molecules under extreme conditions is beginning to emerge. Stromatolite, both fossilized and living, serve as valuable test beds across temporal and spatial scales to expand our search parameters for molecular signatures of life.

4.6 Using Stromatolites to Understand the Constraints of Habitability

For the field of astrobiology, stromatolites serve as valuable resources to explore the intersection between the biosphere and geosphere as it relates to Earth's habitability. Even if stromatolites, as we know them, do not exist beyond Earth, the processes by which they form, including microbe–mineral interactions and the subsequent feedbacks that alter the environment, are likely to occur wherever life is present. In reality, there are no perfect modern analogs for the ancient Earth or extraterrestrial worlds, such as Mars or Europa; however, just as on Earth, the majority of life elsewhere in the universe is expected to be microbial and likely exist as part of a structured ecosystem. Thus, the results from stromatolite research are highly informative for future life detection missions. In particular, studies of modern stromatolites allow the science community to address fundamental questions in astrobiology that go beyond just the detection of life. What is the role of biodiversity in biogeochemical processes and what are the effects of geochemical processes on evolutionary dynamics? How do microorganisms modify and adapt to their abiotic environment and how does that, in turn, alter environmental habitability? What constrains a seemingly "equilibrium" state and what are the forces that produce ecosystem and evolutionary transitions?

The search for life beyond Earth is, therefore, likely to be a search for microbial-based ecosystems, such as stromatolites, and their fingerprints. Understanding the emergent properties of a microbial ecosystem, i.e., those taxa, genes and functions that are important for a particular niche, can have a big impact on our understanding of that environment including how physical and chemical environmental parameters are affected by interacting microbial communities. The co-evolution between microbial communities and their environments result in "ecosystem signatures," from which we can begin to develop generalizable predictions about the habitability of understudied or unknown environments. The study of microbial ecosystems, such as stromatolites, enables us to explore fundamental questions regarding ecosystem functionally to tell the story of life on Earth and, perhaps one day, the universe.

Acknowledgments The authors would like to thank Brooke Vitek for providing an aerial image of the Hamelin Pool. This work was supported in part by a NASA Exobiology and Evolutionary Biology award (NNX14AK14G) to JSF and RPR as well as a NASA NESSF fellowship (80NSSC17K0497) to JB.

References

Allwood AC, Rosing MT, Flannery DT, Hurowitz JA, and Heirwegh CM (2018) Reassessing evidence of life in 3,700-million-year-old rocks of Greenland. Nature 563:241–244
Alleon J, Summons RE (2019) Organic geochemical approaches to understanding early life. Free Radic Biol Med 140:103–112

Andres MS, Reid RP (2006) Growth morphologies of modern marine stromatolites: a case study from Highborne Cay, Bahamas. Sediment Geol 185(3–4):319–328

Andres M, Sumner D, Reid RP, Swart PK (2006) Isotopic fingerprints of microbial respiration in aragonite from Bahamian stromatolites. Geology 34(11):973–976

Aubrey A, Cleaves HJ, Chalmers JH, Skelley AM, Mathies RA, Grunthaner FJ, Ehrenfreund P, Bada JL (2006) Sulfate minerals and organic compounds on Mars. Geology 34(5):357–360

Awramik SM, Margulis L, Barghoorn ES (1976) Evoulutionary processes in the formation of stromatolites. In: Walter MR (ed) Stromatolites. Elsevier, Amsterdam, pp 149–162

Awramik SM (1991) Archaean and Proterozoic stromatolites. In: Calcareous algae and stromatolites, Springer, Berlin, Heidelberg, pp 289–304

Babilonia J, Casaburi G, Louyakis AS, Conesa A, Reid RP, Foster JS (2018) Comparative metagenomics provides insight into the ecosystem functioning of the Shark Bay stromatolites, Western Australia. Front Microbiol 9:1359

Blanchard GF, Paterson DM, Stal LJ, Richard P, Galois R, Huet V, Kelly J, Honeywill C, De Brouwer JFC, Dyer K, Christie M (2000) The effect of geomorphological structures on potential biostabilisation by microphytobenthos on intertidal mudflats. Cont Shelf Res 20(10):1243–1256

Bontognali TRR, Sessions AL, Allwood AC, Fischer WW, Grotzinger JP, Summons RE, Eiler JM (2012) Sulfur isotopes of organic matter preserved in 3.45-billion-year-old stromatolites reveal microbial metabolism. Proc Natl Acad Sci U S A 109(38):15146–15151

Bowlin EM, Klaus J, Foster JS, Andres M, Custals L, Reid RP (2012) Environmental controls on microbial community cycling in modern marine stromatolites. Sediment Geol 263-264:45–55

Braissant O, Decho AW, Dupraz C, Glunk C, Przekop KM, Visscher PT (2007) Exopolymeric substances of sulfate-reducing bacteria: interactions with calcium at alkaline pH and implication for formation of carbonate minerals. Geobiology 5(4):401–411

Breitbart M, Hoare A, Nitti A, Siefert J, Haynes M, Dinsdale E, Edwards R, Souza V, Rohwer F, Hollander D (2009) Metagenomic and stable isotopic analyses of modern freshwater microbialites in Cuatro Ciénegas, Mexico. Environ Microbiol 11(1):16–34

Budd DA, Hajek R, Purkis SJ (2016) Autogenic dynamics and self-organization in sedimentary systems. Soc Sediment Geol 106

Camazine S, Deneubourg JL, Franks NR, Sneyd J, Bonabeau E, Theraula G (2003) Self-organization in biological systems. Princeton University Press, Princeton

Casaburi G, Duscher AA, Reid RP, Foster JS (2016) Characterization of the stromatolite microbiome from Little Darby Island, The Bahamas using predictive and whole shotgun metagenomic analysis. Environ Microbiol 18:1452–1469

Cerqueda-Garcia D, Falcon LI (2016) Metabolic potential of microbial mats and microbialites: autotrophic capabilities described by an in silico stoichiometric approach from shared genomic resources. J Bioinforma Comput Biol 14(4):1650020

Chaçon E (2010) Microbial mats as a source of biosignatures. In: Seckbach J, Oren A (eds) Microbial mats. Springer, Dordrecht, pp 149–181

Chagas AAP, Webb GE, Burne RA, Southam G (2016) Modern lacustrine microbialites: towards a synthesis of aqueous and carbonate geochemistry and mineralogy. Earth Sci Rev 162:338–363

Chan MA, Hinman NW, Potter-McIntyre SL, Schubert KE, Gillams RJ, Awramik SM, Boston PJ, Bower DM, Des Marais DJ, Farmer JD, Jia TZ, King PL, Hazen RM, Leveille RJ, Papineau D, Rempfert KR, Sanchez-Roman M, Spear JR, Southam G, Stern JC, Cleaves HJ (2019) Deciphering biosignatures in planetary contexts. Astrobiology 19(9):1075–1102

Cornet T, Cordier D, Bahers TL, Bourgeois O, Fleurant C, Mouélic SL, Altobelli N (2015) Dissolution on Titan and on Earth: towards the age of Titan's karstic landspaces. J Geophys Res Planets 120(6):1044–1074

Desnues CG, Rodriguez-Brito B, Rayhawk S, Kelley S, Tran T, Haynes M, Lui H, Hall D, Angly FE, Edwards RA, Thurber RV, Reid RP, Siefert J, Souza V, Valentine D, Swan B, Breitbart M, Rohwer F (2008) Biodiversity and biogeography of phages in modern stromatolites and thrombolites. Nature 452:340–345

Dupraz C, Reid RP, Braissant O, Decho AW, Norman RS, Visscher PT (2009) Processes of carbonate precipitation in modern microbial mats. Earth Sci Rev 96(3):141–162

Edgcomb VP, Bernhard JM, Beaudoin D, Pruss S, Welander PV, Schubotz F, Mehay S, Gillespie AL, Summons RE (2013) Molecular indicators of microbial diversity in oolitic sands of Highborne Cay, Bahamas. Geobiology 11(3):234–251

Edgcomb VP, Bernhard JM, Summons RE, Orsi W, Beaudoin D, Visscher PT (2014) Active eukaryotes in microbialites from Highborne Cay, Bahamas, and Hamelin Pool (Shark Bay), Australia. ISME J 8(2):418–429

Flood BE, Bailey JV, Biddle JF (2014) Horizontal gene transfer and the rock record: comparative genomics of phylogenetically distant bacteria that induce wrinkle structure formation in modern sediments. Geobiology 12(2):119–132

Frantz CM, Petryshyn VA, Corsetti FA (2015) Grain trapping by filamentous cyanobacterial and algal mats: implications for stromatolite microfabrics through time. Geobiology 13(5):409–423

Gomez-Acata ES, Centeno CM, Falcon LI (2019) Methods for extracting 'omes from microbialites. J Microbiol Methods 160:1–10

Gorgé O, Bennett EA, Massilani D, Daligault J, Pruvost M, Geigl EM, Grange T (2016) Analysis of ancient DNA in microbial ecology. In: Martin F, Uroz S (eds) Microbial environmental genomics. Methods in molecular biology, vol 1399. Humana Press, New York, pp 289–315

Grotzinger JP, Knoll AH (1999) Stromatolites in Precambrian carbonates: evolutionary mileposts or environmental dipsticks? Annu Rev Earth Planet Sci 27:313–358

Hays LE, Graham HV, Des Marais DJ, Hausrath EM, Horgan B, McCollom TM, Parenteau MN, Potter-McIntyre SL, Williams AJ, and Lynch KL (2017) Biosignature Preservation and Detection in Mars Analog Environments. Astrobiology 17:363–400

Hazen RM, Sverjensky DA (2010) Mineral surfaces, geochemical complexities, and the origins of life. Cold Spring Harb Perspect Biol 2(5):a002162

Jahnert RJ, Collins LB (2012) Characteristics, distribution and morphogenesis of subtidal microbial systems in Shark Bay, Australia. Mar Geol 303-306:115–136

Jahnke LL, Embaye T, Hope JM, Turk KA, Van Zuilen M, Des Marais DJ, Farmer JD, Summons RE (2004) Lipid biomarker and carbon isotopic signatures for stromatolite-forming, microbial mat communities and *Phormidium* cultures from Yellowstone National Park. Geobiology 2(1):31–47

Jahnke LL, Orphan VJ, Embaye T, Turk KA, Kubo MD, Summons RE, Des Marais DJ (2008) Lipid biomarker and phylogenetic analyses to reveal archaeal biodiversity and distribution in hypersaline microbial mat and underlying sediment. Geobiology 6(4):394–410

Kamber BS, Webb GE (2001) The geochemistry of late Archaean microbial carbonate: implications for ocean chemistry and continental erosion history. Geochimica et Cosmochimica Acta 65:2509–2525

Khodadad CL, Foster JS (2012) Metagenomic and metabolic profiling of nonlithifying and lithifying stromatolitic mats of Highborne Cay, The Bahamas. PLoS One 7(5):e38229

Kilburn MR, Wacey D (2011) Elemental and isotopic analysis by NanoSIMS: insights for the study of stromatolites and early life on Earth. In: Stromatolites: interaction of microbe with sediments. Springer, Dordrecht, pp 463–493

Koonin EV, Wolf YI (2012) Evolution of microbes and viruses: a paradigm shift in evolutionary biology? Front Cell Infect Microbiol 2:119

Liu QX, Weerman EJ, Herman PM, Olff H, van de Koppel J (2012) Alternative mechanisms alter the emergent properties of self-organization in mussel beds. Proc Biol Sci 279(1739):2744–2753

Logan BW, Hoffman P, Gebelein CD (1974) Alga mats, cryptalgal fabrics and structures, Hamelin Pool, Western Australia. In: Logan BW (ed) Evolution and diagenesis of quaternary carbonate sequences, Shark Bay, Western Australia, vol 22. American Association of Petrology, Geology and Mineralogy, Tulsa, pp 140–194

Lorenz RD, Mitchell KL, Kirk RL, Hayes AG, Aharonson O, Zebker HA, Pailou P, Radebaugh J, Lunine JI, Janssen MA, Wall SD (2008) Titan's inventory of organic surface materials. Geophys Res Lett 35(2)

Louyakis AS, Mobberley JM, Vitek BE, Visscher PT, Hagan PD, Reid RP, Kozdon R, Orland IJ, Valley JW, Planavsky NJ, Casaburi G, Foster JS (2017) A study of the microbial spatial heterogeneity of Bahamian thrombolites using molecular, biochemical, and stable isotope analyses. Astrobiology 17(5):413–430

Louyakis AS, Gourle H, Casaburi G, Bonjawo RME, Duscher AA, Foster JS (2018) A year in the life of a thrombolite: comparative metatranscriptomics reveals dynamic metabolic changes over diel and seasonal cycles. Environ Microbiol 20(2):842–861

Lowe DR (1980) Stromatolites 3,400-Myr old from the Archaean of Western Australia. Nature 284:441–443

Lund S, Platzman E, Thouveny N, Camoin G, Corsetti F, Berelson W (2010) Biological control of paleomagnetic remanence acquisition in carbonate framework rocks of the Tahiti coral reef. Earth Planet Sci Lett 298(1–2):14–22

Masse M, Bourgeois O, Le Mouelic S, Verpoorter C, Spiga A, Le Deit L (2012) Wide distribution and glacial origin of polar gypsum on Mars. Earth Planet Sci Lett 317:44–55

McCord TB, Hansen GB, Matson DL, Johnson TV, Crowley JK, Fanale FP, Carlson RW, Smythe WD, Martin PD, Hibbitts CA, Granahan JC, Ocampo A (1999) Hydrated salt minerals on Europa's surface from the Galileo near-infrared mapping spectrometer (NIMS) investigation. J Geophys Res Planets 104(E5):11827–11851

Mobberley JM, Ortega MC, Foster JS (2012) Comparative microbial diversity analyses of modern marine thrombolitic mats by barcoded pyrosequencing. Environ Microbiol 14:82–100

Mobberley JM, Khodadad CL, Visscher PT, Reid RP, Hagan P, Foster JS (2015) Inner workings of thrombolites: spatial gradients of metabolic activity as revealed by metatranscriptome profiling. Sci Rep 5:12601

Niles PB, Catling DC, Berger G, Chassefiere E, Ehlmann BL, Michalski JR, Morris R, Ruff SW, Sutter B (2013) Geochemistry of carbonates on Mars: implications for climate history and nature of aqueous environments. Space Sci Rev 174(1–4):301–328

Noffke N, Christian D, Wacey D, Hazen RM (2013) Microbially induced sedimentary structures recording an ancient ecosystem in the ca. 3.48 billion-year-old Dresser Formation, Pilbara, Western Australia. Astrobiology 13(12):1103–1124

Nutman AP, Bennett VC, Friend CR, Van Kranendonk MJ, Rothhacker L, Chivas AR (2019) Cross-examining Earth's oldest stromatolites: seeing through the effects of heterogeneous deformation, metamorphism and metasomatism affecting Isua (Greenlad) ~ 3700 Ma sedimentary rocks. Precambrian Res 331:105347

O'Reilly SS, Mariotti G, Winter AR, Newman SA, Matys ED, McDermott F, Pruss SB, Bosak T, Summons RE, Klepac-Ceraj V (2017) Molecular biosignatures reveal common benthic microbial sources of organic matter in ooids and grapestones from Pigeon Cay, The Bahamas. Geobiology 15(1):112–130

Olendzenski L, Gogarten JP (2009) Evolution of genes and organisms: the tree/web of life in light of horizontal gene transfer. Ann N Y Acad Sci 1178:137–145

Orlando TM, McCord TB, Grieves GA (2005) The chemical nature of Europa surface material and the relation to a subsurface ocean. Icarus 177(2):528–533

Ourisson G, Albrecht P, Rohmer M (1982) Predictive microbial biochemistry - from molecular fossils to procaryotic membranes. Trends Biochem Sci 7(7):236–239

Pace A, Bourillot R, Bouton A, Vennin E, Braissant O, Dupraz C, Duteil T, Bundeleva I, Patrier P, Galaup S, Yokoyama Y, Franceschi M, Virgone A, Visscher PT (2018) Formation of stromatolite lamina at the interface of oxygenic-anoxygenic photosynthesis. Geobiology 16:378–398

Peimbert M, Alcaraz LD, Bonilla-Rosso G, Olmedo-Alvarez G, Garcia-Oliva F, Segovia L, Eguiarte LE, Souza V (2012) Comparative metagenomics of two microbial mats at Cuatro Ciénegas Basin I: ancient lessons on how to cope with an environment under severe nutrient stress. Astrobiology 12(7):648–658

Petryshyn VA, Corsetti FA, Frantz CM, Lund SP (2016) Magnetic susceptibility as a biosignature in stromatolites. Earth Planet Sci Lett 437:66–75

Planavsky N, Reid RP, Andres M, Visscher PT, Myshrall KL, Lyons TW (2009) Formation and diagenesis of modern marine calcified cyanobacteria. Geobiology 7:566–576

Playford PE, Cockbain AE, Berry PF, Roberts AP, Haines PW, Brooke BP (2013) The geology of Shark Bay. Geol Surv West Aust 146:299

Reid RP, Visscher PT, Decho AW, Stolz JF, Bebout BM, Dupraz C, Macintyre IG, Paerl HW, Pinckney JL, Prufert-Bebout L, Steppe TF, DesMarais DJ (2000) The role of microbes in accretion, lamination and early lithification of modern marine stromatolites. Nature 406(6799):989–992

Reid RP, James NP, Macintyre IG, Dupraz CP (2003) Shark Bay stromatolites: microfabrics and reinterpretation of origins. Facies 49:299–324

Reid P, Foster JS, Radtke G, Golubic S (2011) Modern marine stromatolites of Little Darby Island, Exuma archipelago, Bahamas: environmental setting, accretion mechanisms and role of euendoliths. In: Reitner J, Thrauth MH, Stüwe K, Yuen D (eds) Advances in Stromatolite geology. Springer, Berlin, pp 77–90

Reid RP (2011) Stromatolites. In: Hopelys D (ed) Encylopedia of modern coral reefs: structure, form and process. Springer, The Netherlands, pp 1045–1051

Rietkerk M, van de Koppel J (2008) Regular pattern formation in real ecosystems. Trends Ecol Evol 23(3):169–175

Rietkerk M, Dekker SC, de Ruiter PC, van de Koppel J (2004) Self-organized patchiness and catastrophic shifts in ecosystems. Science 305(5692):1926–1929

Roling WF, Aerts JW, Patty CH, ten Kate IL, Ehrenfreund P, Direito SO (2015) The significance of microbe-mineral-biomarker interactions in the detection of life on Mars and beyond. Astrobiology 15(6):492–507

Ruvindy R, White RA III, Neilan BA, Burns BP (2016) Unravelling core microbial metabolisms in the hypersaline microbial mats of Shark Bay using high-throughput metagenomics. ISME J 10:183–196

Saghaï A, Zivanovic Y, Zeyen N, Moreira D, Benzerara K, Deschamps P, Bertolino P, Ragon M, Tavera R, Lopez-Archilla AI, Lopez-Garcia P (2015) Metagenome-based diversity analyses suggest a significant contribution of non-cyanobacterial lineages to carbonate precipitation in modern microbialites. Front Microbiol 6:797

Shapiro B, Hofreiter M (2014) A paleogenomic perspective on evolution and gene function: new insights from ancient DNA. Science 343(6169):1236573

Shen Y, Buick R, Canfield DE (2001) Isotopic evidence for microbial sulphate reduction in the early Archaean era. Nature 410(6824):77–81

Sumner D, and Bowring SA. (1996) U-Pb geochronologic constraints on deposition of the Cambrellrand Subgroup, Transvaal Supergroup, South Africa. Precambrian Research, 79:25–35

Summons RE, Jahnke LL, Hope JM, Logan GA (1999) 2-Methylhopanoids as biomarkers for cyanobacterial oxygenic photosynthesis. Nature 400(6744):554–557

Summons RE, Albrecht P, McDonald G, Moldowan JM (2008) Molecular biosignatures. In: Botta O, Bada JL, Gomez-Elvira J, Javaux E, Selsis F, Summons RE (eds) Strategies of life detection. Space sciences series of ISSI, vol 25. Springer, Boston, pp 133–159

Summons RE, Bird LR, Gillespie AL, Pruss SB, Roberts M, Sessions AL (2013) Lipid biomarkers in ooids from different locations and ages: evidence for a common bacterial flora. Geobiology 11(5):420–436

Suosaari EP, Reid RP, Abreau TA, Playford PE, Holley DK, McNamara KJ, Eberl GP (2016a) Environmental pressures influencing living stromatolites in Hamelin Pool, Shark Bay, Western Australia. PALAIOS 31:483–496

Suosaari EP, Reid RP, Playford PE, Foster JS, Stolz JF, Casaburi G, Hagan PD, Chirayath V, Macintyre IG, Planavsky NJ, Eberli GP (2016b) New multi-scale perspectives on the stromatolites of Shark Bay, Western Australia. Sci Rep 6:20557

Suosaari EP, Reid RP, Oehlert AM, Playford PE, Steffensen CK, Andres MS, Suosaari GV, Milano GR, Eberli GP (2019) Stromatolite provinces of Hamelin Pool: physiographical controls on stromatolites and associated lithofacies. J Sediment Res 89:207–226

Turing AM (1952) The chemical basis of morphogenesis. R Soc Lond B Biol Sci 237(641):37–72

van de Koppel J, Crain CM (2006) Scale-dependent inhibition drives regular tussock spacing in a freshwater marsh. Am Nat 168(5):E136–E147

Van Kranendonk MJ, Webb GE, and Kamber BS (2003) Geological and trace element evidence for a marine sedimentary environment of deposition and biogenicity of 3.45 Ga stromatolitic carbonates in the Pilbara Craton, and support for a reducing Archaean Ocean. Geobiology 1:91–108

Visscher PT, Reid RP, Bebout BM, Hoeft SE, Macintyre IG, Thompson JA (1998) Formation of lithified micritic laminae in modern marine stromatolites (Bahamas): the role of sulfur cycling. Am Mineral 83:1482–1493

Warden JG, Casaburi G, Omelon CR, Bennett PC, Breecker DO, Foster JS (2016) Characterization of microbial mat microbiomes in the modern thrombolite ecosystem of Lake Clifton, Western Australia using shotgun metagenomics. Front Microbiol 7:1064

Walter MR, Buick R, and Dunlop JS (1980) Stromatolites 3,400 - 3,500 Myr old from the North Pole area, Western Australia. Nature 284:443–445

Weber HS, Thamdrup B, Habicht KS (2016) High sulfur isotope fractionation associated with anaerobic oxidation of methane in a low-sulfate, iron-rich environment. Front Earth Sci 4:61

White RA 3rd, Power IM, Dipple GM, Southam G, Suttle CA (2015) Metagenomic analysis reveals that modern microbialites and polar microbial mats have similar taxonomic and functional potential. Front Microbiol 6:966

White RA 3rd, Chan AM, Gavelis GS, Leander BS, Brady AL, Slater GF, Lim DS, Suttle CA (2016) Metagenomic analysis suggests modern freshwater microbialites harbor a distinct core microbial community. Front Microbiol 6:1531

White RA 3rd, Wong HL, Ruvindy R, Neilan BA, Burns BP (2018) Viral communities of Shark Bay modern Stromatolites. Front Microbiol 9:1223

Chapter 5
The Global Distribution of Modern Microbialites: Not So Uncommon After All

Richard Allen White III

Abstract Modern microbialites, a specialized microbial mat that precipitates carbonates, have been thought over the last two decades to be extremely rare, with some in Shark Bay, Western Australia; Highborne Cay, Bahamas; and Cuatro Ciénegas Basin, Mexico. Like their fossilized ancient cousins, the stromatolites represent the oldest evidence for life on Earth (~3.4 Gya), and modern microbialites are prevalent across the globe and have colonized most environments. Many of these microbialite-forming ecosystems have been used as models for astrobiology and NASA mission analogs, including Shark Bay, Pavilion, and Kelly Lakes. This perspective will discuss the global distribution of modern microbialites, their use as astrobiology/NASA mission analogs, and the wealth of information already yielded from 'omics studies obtained from these ecosystems.

5.1 Physical Classification and Brief History

The term microbialite refers to a microbially derived organosedimentary carbonate structure and is thought to be Earth's earliest evidence of life (Grotzinger and Knoll 1999; Schopf 2006; Dupraz et al. 2009). Microbialites are classified into two supergroups according to their macrofabric carbonate layout, as well as gross morphology (Perry et al. 2007; Riding 2011). The first supergroup contains internally laminated carbonate macrofabric stromatolites (thinly layered piles), oncolites (layered, but spherical), and leiolites (amorphous mesofabric) (Perry et al. 2007; Riding 2011; Fig. 5.1). The second supergroup contains unlaminated carbonate macrofabric thrombolites (irregular clots) and dendrolites (branched projections that are tree-like) (Perry et al. 2007; Riding 2011; Fig. 5.1). The classification of microbialites is complicated by their plasticity in nature, with many gradations existing between

R. A. White III (✉)
Plant Pathology, Washington State University, Pullman, WA, United States

RAW Molecular Systems (RMS) LLC, Spokane, Washington, United States

Australian Centre for Astrobiology, University of New South Wales, Sydney, NSW, Australia

© Springer Nature Switzerland AG 2020
V. Souza et al. (eds.), *Astrobiology and Cuatro Ciénegas Basin as an Analog of Early Earth*, Cuatro Ciénegas Basin: An Endangered Hyperdiverse Oasis, https://doi.org/10.1007/978-3-030-46087-7_5

Internal Laminated Macrofabric - Group I

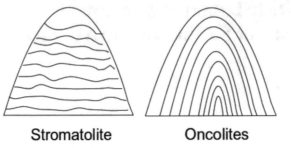

Stromatolite   Oncolites

Outline Non-Laminated Macrofabric - Group II

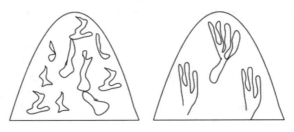

Thrombolite   Dendrolite

Amorphous with aphanitic Macrofabric - Group III

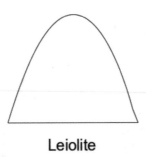

Leiolite

Fig. 5.1 Microbialite morphology and carbonate macrofabric layout. (Adapted from Riding 2011)

these basic morphologies. Among living forms, stromatolites and thrombolites are the most common (Dupraz et al. 2009; Fig. 5.2).

Microbialites have been intensely studied for over one hundred years (Kalkowsky 1908). The first mention of the term stromatolite, or "stromatolith," was used to describe early Triassic stromatolites (Kalkowsky 1908; Riding 2011; Fig. 5.1). Pia (1933) was the first to classify oncolites as "unattached, regularly or irregularly spheroidal, concentrically to semiconcentrically laminated bodies" (Fig. 5.1). Oncolites were reclassified as "a category of stromatolite structure," thus removing

Fig. 5.2 Microbialite morphology examples. (**a**) A fossilized stromatolite (~505 Mya). Burgess shale in British Columbia (Photo courtesy of Dr. David Turner, Beatty Biodiversity with permission). Gray scale bar 7 cm. (**b**) A close-up of the fossilized stromatolite (10× from scale bar). (**c**) Clinton Creek thrombolite from Clinton Creek, Yukon Canada (Photo courtesy of Dr. Ian Malcolm Power (University of British Columbia) with permission). Gray scale bar 3 cm. (**d**) A close up of the Clinton Creek thrombolite (10× from scale bar)

them as a separate category (Pia 1933; Logan et al. 1964). Aitken (1967) was the first to use the term "thrombolite" from the Greek meaning "thrombos or blood clot," from their irregular, clot-like arrangement (Fig. 5.1). Burne and Moore (1987) effectively provided the modern definition we use today, as structures "that are accreted as a result of a benthic microbial community trapping and binding detrital sediment and/or forming the locus of mineral precipitation." The definition was broadened from one that included biotic factors to that of a "benthic microbial community," took into account abiotic factors, such as "detrital sediment," and added geologic context ("forming the locus of mineral precipitation").

5.2 Microbialites as the Oldest Persistent Ecosystems

Microbialites represent the oldest known persistent ecosystems and potentially the earliest evidence of life (Schopf 2006; Grotzinger and Knoll 1999). The geologic record indicates that microbialite diversification occurred concurrently with the emergence of life ~3.50 billion years ago (Gyr), followed by a period of dominance extending until ~542 million years ago (Mya) following the Cambrian explosion (Allwood et al. 2006; Dupraz and Visscher 2005; Schopf 2006; Rothschild and Mancinelli 1990; Fig. 5.3). With the age of the Earth estimated to be ~4.54 billion years (Gyr), the appearance of microbialites in the fossil record at ~3.85–3.5 Gyr suggests that microbialites have persisted for ~85% of geological history (Patterson 1956; Dupraz and Visscher 2005; Fig. 5.3). Modern microbialites first appear when

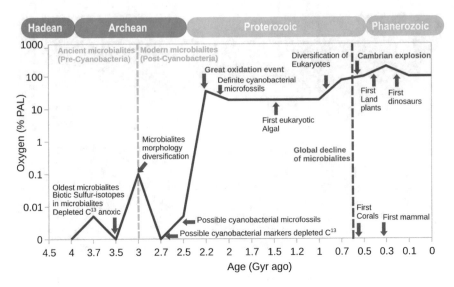

Fig. 5.3 Timeline of microbialite and global atmospheric oxygen concentration over geological time. (The figure was adapted from Canfield 2005 and Dupraz and Visscher 2005. The oxygen data were provided by Dr. Sean Crowe (University of British Columbia) with permission, including data from Lyons et al. 2014 and Crowe et al. 2013. Oxygen data are expressed as %PAL or present atmospheric levels)

cyanobacteria emerge (~3.0 Gyr), persisting for roughly 66% of Earth's geologic history (Dupraz and Visscher 2005; Fig. 5.3). Thus, modern microbialites are the best available representation of Earth's earliest life forms, and their biology may grant insight into the oldest aspects of the biosphere (Grotzinger and Knoll 1999; Dupraz and Visscher 2005).

5.2.1 Global Distribution of Microbialites

Microbialites have adapted to radiate into many environmental conditions, including extremes of temperature, pH, and salinity (Dupraz and Visscher 2005). Microbialites are globally distributed and can be found in the open ocean (Reid et al. 2000; Khodadad and Foster 2012; Mobberley et al. 2013), freshwaters (Ferris et al. 1997; Laval et al. 2000; Gischler et al. 2008; Breitbart et al. 2009), hypersaline lagoons (Goh et al. 2009; Allen et al. 2009), tropical lagoons (Sprachta et al. 2001), hot springs (Bosak et al. 2012), and remnant mining sites (Power et al. 2011a, b). To a lesser extent, they have also colonized terrestrial environments, such as soils (Maliva et al. 2000) and caves (Lundberg and Mcfalane 2011).

Modern microbialite-forming ecosystems have been found globally, suggesting that they are not as rare as previously thought. North America and Antarctica harbor many examples of freshwater microbialites, with Cuatro Ciénegas and Yellowstone microbialites being the most studied (Breitbart et al. 2009; Bosak et al. 2012). In North America, microbialites are found in the Yukon (Power et al. 2011a), Southwestern British Columbia (Laval et al. 2000; Ferris et al. 1997), Yellowstone National Park in Wyoming (Bosak et al. 2012), Pyramid Lake Nevada (Arp et al. 1999), and in Mexico at Cuatro Ciénegas (Breitbart et al. 2009), Lake Pátzcuaro (Bridgwater et al. 1999; Bischoff et al. 2004), and Lake Alchichica (Couradeau et al. 2011). Antarctica is a haven for microbialites, with many found in pristine condition with little human influence. Antarctica has four ecosystems containing conical stromatolites or thrombolites: Lake Untersee (Andersen et al. 2011), Lake Joyce (Hawes et al. 2011), Lake Hoare (Squyres et al. 1991), and Lake Vanda (Andersen et al. 2011).

The microbialites of Central and South America are found in some of the most extreme environments, including lakes containing toxic levels of heavy metals. Microbialites are found in Central and Mesoamerica, including Laguna Bacalar, Mexico, which borders with Belize (Gischler et al. 2008). In South America, microbialites are found in Brazil in the hypersaline lakes of Lagoa Vermelha and Lagoa Salgada (Vasconcelos et al. 2006; Spadafora et al. 2010), in northern Chile in La Brava (Farías et al. 2012), and at the southern tip of Chile (Graham et al. 2014). Laguna Socompa on the border of Chile and Argentina is an example of an extreme ecosystem that contains microbialites, as it contains high levels of arsenic (18.5 mg L^{-1}) and iron (1 mg L^{-1}) (Farías et al. 2012).

Diverse microbialites, including rare hydromagnesite forms and the largest microbialite ever recorded, are found in Europe. The Ruidera pools in Spain hold a great diversity of freshwater stromatolites (Santos et al. 2010), and Lake Van, Turkey, contains the largest microbialites on record (Kempe et al. 1991; López-García et al. 2005). Salda Gölü Lake, in southern Turkey, also holds the only other example of hydromagnesite microbialites outside of Lake Alchichica, Mexico (Braithwaite and Zedefa 1994; Couradeau et al. 2011).

Microbialites in Oceania are rare; however, they include modern examples of coral replacement. Hypersaline lakes found in Lake Stonada, Indonesia (Arp et al. 2004), Kiritimati Atoll in the central Pacific (Schneider et al. 2013), and the caldera lakes of Niuafo'ou Island, Tonga (Kazmierczak and Kempe 2006) contain microbialites. Microbialites in the tropical lagoons of French Polynesia and New Caledonia exist side by side with dead and live corals, illustrating direct competition (Sprachta et al. 2001).

Western Australia has the oldest fossilized examples of microbialites at roughly 3.45 Gyr old (Van Kranendonk et al. 2008), and Shark Bay harbor is one of the few examples of marine stromatolites (Goh et al. 2009; Allen et al. 2009). Two lakes in Australia have documented microbialites, including Lake Thetis (Grey et al. 1990) and Lake Clifton (Konishi et al. 2001; Warden et al. 2016). Other lakes in Australia that contain thrombolitic microbialites include Lake Richmond, Lake Walyungup, and Lake Cooloongup, which are protected and endangered; however, no publications exist describing them.

While China and Russia, the largest combined landmass on the planet, have many examples of microbialite fossils, no studies describe modern versions, or if studies are available, they have not been translated into English (Wu et al. 2014). In terms of African microbialites, a recent study found some in southern Africa (Smith et al. 2011), and fossils have been found throughout the continent. However, Africa remains largely unstudied when it comes to modern microbialites.

5.2.2 Proposed Microbialite Formation Hypotheses

There is much debate on whether the formation of ancient microbialites dating back ~3.4 Gya in Western Australia was purely abiotic (Lowe 1994; Grotzinger and Rothman 1996) or a combined biotic–abiotic event (Dupraz and Visscher 2005; Schopf 2006; McLoughlin et al. 2008; Dupraz et al. 2009).

5.2.2.1 Abiotic (Spectator Hypothesis)

The abiotic model stresses that ancient microbialites and possibly even modern versions are formed completely by abiotic processes (Lowe 1994; Grotzinger and Rothman 1996; McLoughlin et al. 2008). Lowe suggested that the Western Australian stromatolites within the Warrawoona group (3.5–3.2 Gya) lack the structural evidence of biotic processes and therefore are of abiotic origin (1994). Moreover, a laboratory spray deposition experiment showed that stromatolites could be generated abiotically using enamel spray paint with metal-oxide pigment particles (McLoughlin et al. 2008). A purely abiotic model would suggest that the microbes and/or pathways would be the same between the microbialite and their surrounding environment, as the chemistry alone is driving the formation of the microbialite: microbes would merely be "spectators." Thus, the abiotic model would suggest that microbes have no role in the formation of microbialites.

Abiotic factors are certainly critical to microbialite formation. These include hard water (≥ 121 mg $CaCO_3$ L^{-1}) with free divalent cations (mainly Ca^{2+} and/or Mg^{2+}), high alkalinity (≥ 121 mg $CaCO_3$ L^{-1}), stable alkaline pH water (pH $\geq$ 8), clear water with visible light penetration, and dissolved inorganic carbon (DIC) (Dupraz et al. 2009). No carbonate-derived microbialite-forming ecosystem has been found without at least hard water, high alkalinity, and a stable alkaline pH (pH $\geq$ 8); carbonates cannot form without these factors.

5.2.2.2 Biotic-Driven Hypothesis

The biotic link to the origin of microbialites has been debated for over one hundred years since it was first suggested by Kalkowsky (1908). Lowe (1994) rejected any biotic role in the formation of the Australian Warrawoona group stromatolites

(3.5–3.2 Gya). Buick et al. (1995) responded that Lowe violated Descartes' fourth rule of science, "omettre rien," directly translating to "omit nothing," suggesting that Lowe (1994) omitted critical examples relating to a combined hypothesis. The examples not included by Lowe (1994) were the post-Isua $\delta^{13}C$ isotopic data, and various microfossil examples illustrating that biotic–abiotic processes are involved in the formation of microbialites (Buick et al. 1995). Since that time, sulfur isotopic measurements in the Australian stromatolites from ~3.45 Gya suggest negative $\delta^{34}S$ isotopic fractionation, which can only be explained by biotic processes (Bontognali et al. 2012, Fig. 5.3). Ancient microbialites spanning ~3.5–3.0 Gya formed under anoxygenic processes, as sufficient oxygen was not present to drive oxygenic photosynthesis, whether cyanobacteria were present or not (Crowe et al. 2013; Planavsky et al. 2014; Fig. 5.3).

Biotic metabolism in modern microbialites is an active and highly complex interaction with abiotic factors. Both oxygenic and anoxygenic metabolisms within the surface microbial mat on microbialites are highly productive (5000–6200 mg biomass cm^{-2} day^{-1}) and similar to the primary productivity of tropical forests (6000 mg biomass cm^{-2} day^{-1}) (Stal 2000; Centeno et al. 2012). Carbonates and free calcium are initially trapped by adhesive biofilms and mats that are often cyanobacterial in origin (Dupraz and Visscher 2005; Dupraz et al. 2009). The cyanobacterial mat is then degraded by heterotrophic bacteria that liberate HCO_3^- and Ca^{2+}, thereby increasing the saturation index and producing nucleation points for carbonate precipitation (Dupraz and Visscher 2005; Dupraz et al. 2009). Cyanobacterial mats contain exopolymeric substances (EPS) that serve as a location of mineral nucleation, while providing a heterotrophic microenvironment favorable for organomineralization via dissimilatory sulfate reduction (Dupraz and Visscher 2005; Dupraz et al. 2009). Light is another critical factor in the biotic–abiotic relationship of microbialite formation. Many microbialite-forming ecosystems are either in shallow locations or have clear light visibility with high light penetration allowing for photoautotrophic microbial mat growth.

The concept that microbes have no role in microbialite formation does not account for evidence that specific mineral precipitation is a function of microbial presence. For the abiotic hypothesis, the question is, why would only certain minerals form, and rates of precipitation differ, with the presence of microbes under similar abiotic conditions? The Atlin wetland in British Columbia, Canada, contains carbonate-precipitating microbial mats. A microcosm experiment showed that abiotic–biotic microcosms promoted carbonate formation at a faster rate and resulted in a completely different mineral (dypingite) than the abiotic microcosm (nesquehonite) (Power et al. 2007). This suggests that both biotic and abiotic forces are required to form particular minerals associated with lithifying microbial mats, including microbialites (Power et al. 2007). Microbialite formation is likely a consequence of complex interactions between both biotic and abiotic factors (Dupraz et al. 2009). However, much is still unknown regarding the effect of microbial community interactions on the formation of microbialites.

5.2.3 Major Members of the Microbialite Microbiome

Microbialite-forming ecosystems, which include microbial mats, contain highly diverse microbial consortia, including both cellular (bacterial, archaea, and eukaryotes) and viral (capsid based) organisms.

5.2.3.1 Alphaproteobacteria

Alphaproteobacteria are abundant in marine, freshwater, and artificial marine-derived microbialite-forming systems (Havemann and Foster 2008; Breitbart et al. 2009; Goh et al. 2009; Khodadad and Foster 2012 ; White III and Suttle 2013) and cyanobacterial nonlithifying mats (Varin et al. 2012). Pozas Azules II within Cuatro Ciénegas and the high-altitude, hypersaline arsenic-rich stromatolites of Lake Scompa within the Argentinean Andes have a high percentage of Alphaproteobacteria sequences (~12–15%), mainly belonging to the Rhodobacterales (Breitbart et al. 2009; Farías et al. 2012). The rare hydromagnesite microbialites of Lake Alchichica and the oncolites of Rios Mesquites had cyanobacterial sequences that outnumber the Alphaproteobacteria sequences (Breitbart et al. 2009; Couradeau et al. 2011).

Alphaproteobacteria contain members (e.g., Rhodobacterales) that are photoheterotrophic, which could influence the formation of microbialites by anoxygenic photosynthesis and microbialite dissolution through fermentation (Dupraz and Visscher 2005; Fig. 5.4). Anoxygenic photosynthesis would been the only type of photosynthesis within the Precambrian stromatolites prior to the origin of cyanobacteria, and alphaproteobacterial anoxygenic phototrophs could have filled that metabolic niche (Bosak et al. 2007). Alphaproteobacteria appear to be important for microbialite nitrogen fixation, even in the presence of cyanobacterial nitrogen fixation, likely because of diel cycles (Havemann and Foster 2008; Fig. 5.4).

5.2.3.2 Planktonic Cyanobacteria

Cyanobacteria in microbialites exist as planktonic cells and mat builders. The planktonic unicellular cyanobacteria are found in the surrounding water. Cyanobacterial calcification is well documented (Dittricha et al. 2003; Obst et al. 2009). *Synechococcus* sp., a model planktonic cyanobacterium, has been shown in multiple studies to precipitate carbonates using oxygenic photosynthesis (Thompson and Ferris 1990; Dittricha et al. 2003; Obst et al. 2009; Fig. 5.4). *Synechococcus* sp. is known to cause "whiting events," in which calcium carbonate rains down through the water column (Thompson et al. 1997). These "whiting events" are a photosynthetically driven alkalization process that precipitates calcium carbonate principally around the *Synechococcus* cells, which serve as nucleation points of precipitation (Dittricha et al. 2003; Obst et al. 2009). These planktonic cyanobacteria provide the

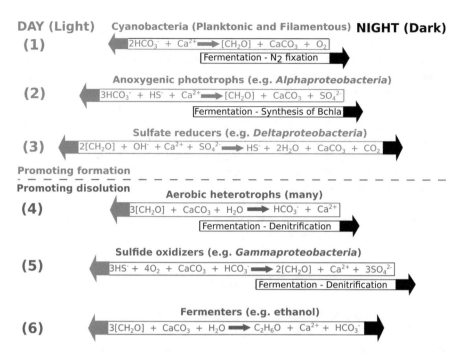

Fig. 5.4 Generalized equations for microbialite formation and dissolution. Includes examples of key taxa involved or substrates (ethanol only) during day (light) and night (dark) cycles. (White III et al. 2016)

"bricks," in the form of either magnesium- or calcium-based carbonates, which are the building blocks of microbialites (Jansson and Northern 2010).

5.2.3.3   Filamentous Cyanobacteria, Mat Builders

If the planktonic cyanobacteria provide the "bricks" to make microbialites, then mat-building filamentous cyanobacteria provide the "mortar" that stabilizes the structure. Filamentous cyanobacteria are fundamental in modern microbialites, as they provide nutrients and structure and directly act in the formation processes. Filamentous cyanobacteria, like their planktonic counterparts, influence carbonate precipitation through photosynthesis-induced alkalinization (Dupraz and Visscher 2005; Fig. 5.4). Mat-building cyanobacteria produce extracellular polymeric substances (EPS), which trap Ca^{2+} ions near dissolved CO_3^{2-}, and provide nucleation points for precipitation (Dupraz and Visscher 2005; Dupraz et al. 2009). Commonly found genera of filamentous mat-building cyanobacteria in microbialite-forming sites include *Nostoc*, *Pseudanabaena*, *Schizothrix*, *Dichothrix*, and *Lyngbya*, (Laval et al. 2000; Dupraz et al. 2009).

Mat-building filamentous cyanobacteria also provide an essential energy source for heterotrophic bacteria in an oligotrophic water column (Lim et al. 2009). These heterotrophic bacteria in turn mediate further lithification and buildup of the microbialite structure (Omelon et al. 2013). Heterotrophs release calcium and/or magnesium ions that are bound to cyanobacterial EPS mat by degradation (Dupraz and Visscher 2005). This process allows for more free interaction between cations and CO_3^{2-} anions and leads to carbonate precipitation (Dupraz and Visscher 2005). Phosphorus, nitrogen, and various vitamins and cofactors are critical nutrients required for primary production by filamentous cyanobacteria; these nutrients can be sequestered in the mats and then released by heterotrophic degradation, which feeds further primary production. These cycles of autotrophic and heterotrophic mat breakdown are balanced; otherwise, the structures would likely dissolve and the mats would die. Through primary production, the microbialite cyanobacterial mat is an oasis in a desert-like oligotrophic water column.

5.2.3.4 Sulfur-Oxidizing Bacteria (SOBs)

Sulfur-oxidizing bacteria (SOBs) are a double-edged sword for microbialite formation, as they directly cause dissolution of microbialites, but byproducts of this dissolution provide intermediate substrates that allow for colonization by sulfate-reducing bacteria (SRBs; Fig. 5.4). Microbialites are often found in highly oxygenic water (Dupraz et al. 2009), and SOBs remove high amounts of oxygen, allowing for further lithification via sulfate reduction by anaerobic sulfate-reducing bacteria (Dupraz and Visscher 2005). SOBs are chemolithotrophs that promote dissolution of microbialites through the oxidation of hydrogen sulfide, which in turn releases calcium ions from the microbialite calcium carbonate matrix (Fig. 5.4).

The sulfur oxidation pathway encodes the *sox* gene cluster that is responsible for aerobic sulfur oxidization (Friedrich et al. 2005). Sox genes are conserved across Proteobacteria, including Gamma-, Beta-, and Alphaproteobacteria (Friedrich et al. 2005). *Beggiatoa* sp., a ubiquitous filamentous sulfide-oxidizing gammaproteobacterium, would be an excellent candidate for the study of the removal of oxygen from the highly oxygenated surface mat of microbialites (Dupraz and Visscher 2005). The *sox* pathway encodes four protein complexes SoxYZ, SoxAX, SoxB, and SoxCD, which oxidize reduced sulfur species, including sulfide (HS⁻), elemental sulfur (S), and sulfite (HSO_3^-) (Friedrich et al. 2005). The diversity and abundance of SOBs, and whether Sox-related pathways are found in active microbialite-forming ecosystems, are still unclear. It is also unclear whether sulfur oxidation pathways are a signature of healthy (actively forming) or failing (actively dissolving) microbialites.

5.2.3.5   Sulfate-Reducing Bacteria (SRBs)

Sulfate-reducing bacteria (SRBs) are a crucial part of the heterotrophic microbial consortia responsible for lithification in microbialites (Visscher et al. 2000; Dupraz and Visscher 2005; Gallagher et al. 2012; Fig. 5.4). Sulfate reduction is an ancient metabolism, predating oxygenic photosynthesis (Canfield et al. 2000). Sulfate reduction was occurring in microbialites prior to the great oxidation event and the appearance of cyanobacterial mats (Canfield et al. 2000; Wacey et al. 2011; Bontognali et al. 2012; Fig. 5.3).

SRBs reduce sulfate under anaerobic conditions, but also exist in active layers of highly oxygenated mats, which presents a metabolic paradox (Dupraz and Visscher 2005; Dupraz et al. 2009). SRBs in microbialites must be oxygen tolerant, find anoxic microenvironments, or form tight consortia with oxygen-depleting anoxic organisms, such as SOBs (Dupraz and Visscher 2005). Recent rate measurements suggest that SRBs within microbialites must be oxygen tolerant, as measurable amounts of dissimilatory sulfate reduction occur in the highly oxygenated surface layers of the cyanobacterial mat (Baumgartner et al. 2006; Gallagher et al. 2012).

SRBs may mediate carbonate precipitation in microbialites by altering the saturation index and increasing alkalinity though dissimilatory sulfate reduction (Baumgartner et al. 2006; Gallagher et al. 2012; Fig. 5.4). Active microbialite growth is thus suggested to be related to the heterotrophic breakdown of cyanobacterial-derived mats, which provide a rich and complex carbon source (Dupraz and Visscher 2005). SRBs need electron donors for functional dissimilatory sulfate reduction: these include organic carbon, as acetate, lactate, ethanol, or formate, or molecular hydrogen (Baumgartner et al. 2006, 2009; Gallagher et al. 2012). The concept of lithifying mats and nonlithifying mats could be explained by the rates of dissimilatory sulfate reduction and related to the abundance of metabolically active SRBs (Dupraz and Visscher 2005; Gallagher et al. 2012). However, this process and the underlying mechanism are not well understood.

SRBs that use the highly conserved dissimilatory sulfate reduction (or *dsr*) pathway are mainly found among Proteobacteria, but are found also in nonproteobacterial lineages. Specifically, the majority of SRB-related taxa are within the class Deltaproteobacteria and fall within the orders Desulfovibrionales, Desulfobacterales, and Syntrophobacterales. Nonproteobacterial lineages that have SRBs include the taxa Nitrospira, Thermodesulfobacteria, Thermodesulfobiaceae, and Firmicutes (mainly Clostridia). The *dsr* pathway consists of sulfate adenylyltransferase (*sat*), adenylylsulfate reductase with the A and B subunit (*apsAB*), and dissimilatory sulfite reductase with alpha and beta subunits (*dsrAB*) (Zhou et al. 2011). The pathway transports sulfate (reactant) with an electron donor (organic carbon, acetate, lactate, formate, hydrogen, and/or ethanol), and through various enzymatic steps, it yields sulfide (Zhou et al. 2011). The diversity and occurrence of the *dsr* pathway in microbialites is not known, but it is critical to our understanding of microbialite formation (Baumgartner et al. 2006; Gallagher et al. 2012).

5.2.3.6 Firmicutes

Firmicutes are Gram-positive bacteria that are abundant in a multitude of environments, including microbialites (Breitbart et al. 2009). *Bacillus* spp. were shown over 40 years ago to produce carbonates as a "general phenomenon" (Boquet et al. 1973). Firmicutes are known to precipitate calcium carbonate with or without the presence of cyanobacteria, with at least one operon (*lcfA*) geared toward carbonate precipitation (Barabesi et al. 2007).

Firmicutes have many mechanisms for facilitating carbonate precipitation. Seven different *Bacillus* species are known to precipitate carbonates, including *B. subtilis* (Barabesi et al. 2007), *B. sphaericus* and *B. lentus* (Dick et al. 2006), *B. megaterium*, *B. cereus*, and *B. thuringiensis* (Dhami et al. 2013), and *B. amyloliquefaciens* (Lee 2003). *Bacillus* spp. are able to induce carbonate formation through the metabolism of urea (Hammes et al. 2003), heterotrophic ammonification of amino acids (Castanier et al. 1999), carbonic anhydrase (Dhami et al. 2014), EPS formation (Ercole et al. 2007), and the *lcfA* operon (Barabesi et al. 2007). Even at low abundance, members of the Firmicutes have metabolic versatility geared toward carbonate precipitation, which may have a role in the formation of microbialites (Paerl et al. 2001; Breitbart et al. 2009; Omelon et al. 2013).

Microbial mats including microbialites have colored pigmented layers (orange to yellow) (Bottos et al. 2008). Little is known about nonphotosynthetic pigmented heterotrophic bacteria in microbialites, although many studies have focused on the oxygenic phototrophs in these layers and suggested that the coloration is related to carotenoid pathways (Nübel et al. 1999; Lionard et al. 2012). Diversity studies (based on 16S rDNA) have found firmicutes in these layers (Bottos et al. 2008; Lionard et al. 2012), and firmicutes synthesize carotenoids that give rise to many different colors of bacteria, including orange (Köcher et al. 2009; Klassen 2010). A recent study has screened for cultivatable isolates of nonphotosynthetic, pigmented, heterotrophic members of the Firmicutes, which found isolates relating to the *Exiguobacterium* genus (White et al. 2013b, 2019). The *Exiguobacterium* genus appears to found in many microbialite-forming sites including Pavilion Lake (White et al. 2019) and Cuarto Ciénegas (Rebollar et al. 2012). The functional roles of these nonphotosynthetic, pigmented, heterotrophic Exiguobacteria are unknown, and they may function in detoxification of heavy metals in microbialite-forming ecosystems (White et al. 2019).

5.2.3.7 Actinobacteria

Representatives of the phylum Actinobacteria appear to be rare within microbialite-forming ecosystems, and little is known about their abundance, diversity, and role. Actinobacteria have been found in smooth and pustular pre-stromatolitic mats in Shark Bay and the freshwater stromatolites of Ruidera Pools, Spain (Allen et al. 2009; Santos et al. 2010). Their presence in microbialites does not entail a functional role, but the knowledge gap points to a need for both culture-dependent and culture-independent studies.

Actinobacteria have also been identified in the pigmented layers in microbial mats (Bottos et al. 2008; Lionard et al. 2012), and it has been suggested that they may be responsible for the coloration (Klassen 2010). Although carotenoid biosynthesis also occurs in Actinobacteria, little work has been completed on nonphotosynthetic pigmented heterotrophic members of the Actinobacteria to determine if they may contribute to the colored layers found in microbial mats and microbialites. Recently, a paper describing an *Agrococcus* (White III et al. 2013a, b, 2018) suggests that it may be involved detoxification of heavy metals.

5.2.3.8 Eukaryotes

Eukaryotes in microbialites are present in low abundance (<1%), by physical counts (Power et al. 2011a) and by sequence abundance (Couradeau et al. 2011; Khodadad and Foster 2012). Even at low abundance, metazoans have been associated with the global decline of microbialites (Awramik 1971; Garrett 1970, Figure 1.3). Metazoan grazing following the Cambrian explosion has been correlated to a drop in stromatolite diversity about ~500–800 Mya (Awramik 1971; Garrett 1970; Fig. 5.3). However, the metazoan grazing hypothesis is hindered by a lack of fossil evidence; as well, protist grazers presumably predated the Cambrian microbialites far earlier (Bernhard et al. 2013). Another major problem is that unicellular or soft body grazers would leave no obvious fossils, possibly contributing to a virtually nonexistent fossil record (Bernhard et al. 2013).

The question remains, why do modern microbialites exist if grazing is the cause of their decline? This is puzzling, as metazoan grazing occurs even under the oligotrophic conditions in which many microbialites are found (Elser et al. 2005). Possibly, metazoan grazing is controlled by abiotic factors, such as low nutrients, high temperature, high salinity, and high pH. Cyanobacteria make cytotoxic compounds that can inhibit and kill metazoans, thus cyanobacterial mats in microbialites could also biologically control metazoan grazing (Neilan et al. 2013; Wu et al. 2014). The type of microbial mat that exists on the surface of a marine microbialite is suggested both to alter the microbialite fabric (whether stromatolitic or thrombolitic) and to control the metazoan community structure (Tarhan et al. 2013). Thus, the microbial community also likely plays a role in controlling metazoan grazing (Tarhan et al. 2013).

In contrast, photoautotrophic eukaryotes (e.g., algae and other protists) could aid in the organosedimentary formation process via photosynthesis-induced carbonate precipitation (Fig. 5.4). Even at low sequence abundances of eukaryotes (<1%) in marine microbialite ecosystems (Highborne Cay, Bahamas), it is suggested that foraminifera influence the microfabric by stabilizing carbonate grains, and their reticulopods, once encased, act as sheaths that effect the lamination of microbialites (Bernhard et al. 2013; Edgcomb et al. 2013). Thus, foraminifera may help to explain the changes in the microbialite microfabric observed in the late Precambrian

(Bernhard et al. 2013). In modern microbialites, foraminifera are suggested to move carbonate grains during diel vertical migration, disrupting lamination critical for stromatolite formation, and promote clot formation or thrombolitic structure (Bernhard et al. 2013). Foraminifera were found in microbialites in the Highborne Cay, Bahamas, including thrombolites and stromatolites; however, the highest diversity was observed in thrombolite mats or mats that were biologically turbated (Bernhard et al. 2013; Edgcomb et al. 2013). Metazoans, Stramenopiles, Alveolata, Amoebozoa, and Rhizaria were found to be active metabolically in various microbialites across Highborne Cay and Shark Bay (Edgcomb et al. 2013).

5.2.3.9   Viruses

The role of viruses in microbialite communities has remained elusive. Viruses are the most prevalent "organisms" on Earth, with an estimated 10^{30} viruses in the ocean and around 10^6–10^7 viruses ml^{-1} in most freshwater systems (Suttle 2005). Viruses are agents of cell lysis and thus play a role in carbon cycling on a global scale (Suttle 2005). Additionally, new research has found that many viruses harbor auxiliary metabolic genes, including genes that encode the core proteins involved in photosynthesis (*psbA*, *psbD*) and phosphate starvation (*phoH*) (Clokie et al. 2010; Thompson et al. 2011). Viral auxiliary metabolic genes have also been shown to redirect their host's metabolic potential to fuel viral replication (Clokie et al. 2010; Thompson et al. 2011). Some marine cyanophages encode a Calvin cycle inhibitor (cp12) that inhibits the carbon-fixation actions of Rubisco (Ribulose-1,5-bisphosphate carboxylase/oxygenase), and a viral-encoded transaldolase (*talC*) redirects the host metabolism to the pentose phosphate pathway, benefiting viral replication (Clokie et al. 2010; Thompson et al. 2011). This evidence suggests that viruses affect the carbon metabolism of an ecosystem and may influence the chemistry of microbialite formation or dissolution.

5.2.4   Previously Studied Microbialite-Forming Ecosystems

Shark Bay in Western Australia, Highborne Cay in Bahamas, and the Cuatro Ciénegas basin of Mexico, including Pozas Azules II and Rios Mesquites, are the most studied microbialite-forming assemblages. The best studied for microbiome studies in freshwater include Pavilion and Kelly Lake and Clinton Creek.

5.2.4.1   Shark Bay Microbialites

The oldest fossil microbialites are found in Western Australia, dating back ~3.5 Gya (Van Kranendonk et al. 2008). Near these fossil microbialites are active, modern versions within Shark Bay at Hamelin Pool in Western Australia. These are the main example of hypersaline marine microbialites (Goh et al. 2009; Allen et al. 2009; Suosaari et al. 2016). Shark Bay microbialites range in morphology from columnar, club, spheroidal, domal, nodular to irregular-shaped morphotypes (Burns et al. 2004).

Shark Bay microbialite diversity has been assessed by three 16S rDNA amplicon studies (Burns et al. 2004; Goh et al. 2009; Allen et al. 2009). Currently, only 412 rDNA sequences are available on GenBank from all bacterial surveys of the Shark Bay microbialites. These sequences reveal the distribution of the various bacterial taxa, indicating that Alphaproteobacteria (20%), Actinobacteria (16%), and Cyanobacteria (15%) dominate, with the rest of the sequences belonging to Planctomycetes (11%), Gammaproteobacteria (10%), and the Firmicutes (9%) (Burns et al. 2004; Papineau et al. 2005; Allen et al. 2009; Goh et al. 2009). Papineau et al. (2005) concluded that Shark Bay microbialites are highly diverse, with 90% of sequences belonging to bacteria, and that Shark Bay microbialites have novel Proteobacteria representing 28% of the sequences, followed by Planctomycetes (17%) and Actinobacteria (14%), respectively.

Shark Bay is novel compared to other microbialite-forming ecosystems as it has a relatively high abundance of archaea (~10% of 16S rDNA sequence), which are usually <1% of the microbial community in other microbialites (Burns et al. 2004; Papineau et al. 2005; Allen et al. 2009; Goh et al. 2009). The archaea in Shark Bay microbialites are principally halophiles (Papineau et al. 2005; Allen et al. 2009; Goh et al. 2009). Many unclassified haloarchaea have been detected only in Shark Bay microbialites and not in the surrounding seawater, indicating that Shark Bay microbialites are a potential reservoir for these archaea (Papineau et al. 2005; Allen et al. 2009; Goh et al. 2009). Metagenomic data are beginning to emerge for Shark Bay microbialites (Babilonia et al. 2018; Ruvindy et al. 2016; Wong et al. 2012) to complement the amplicon libraries (Burns et al. 2004; Papineau et al. 2005; Goh et al. 2009; Wong et al. 2015, 2017), and a RNA-based study on active eukaryotes (Edgcomb et al. 2013) is available for comparison. Recently, a virome for Shark Bay has also been published, describing novel ssDNA viral genomes (White III et al. 2018).

5.2.4.2 Highborne Cay Microbialites

Like Shark Bay, Highborne Cay microbialites are found in shallow depths. They have a high diversity of colored mats covering the microbialites, ranging pink, green, brown, and black (Reid et al. 2000; Myshrall et al. 2010; Mobberley et al. 2013). Highborne Cay stromatolite-forming mats have been classified into three types (I–III) based on morphology (Reid et al. 2000). Type I mats are considered young and are dominated by filamentous cyanobacteria with few heterotrophic bacteria. They have low diversity (~128 OTUs) and low sulfate reduction, photosynthesis, and aerobic respiration rates and have been described as nonlithifying mats (Reid et al. 2000; Myshrall et al. 2010; Khodadad and Foster 2012; Mobberley et al. 2013). Type II and III mats are considered older lithifying mats with the beginning of carbonate lamination; both have higher rates of sulfate reduction, photosynthesis, and aerobic respiration rates than type I mats (Reid et al. 2000; Myshrall et al. 2010;

Khodadad and Foster 2012; Mobberley et al. 2013). Type II and III mats have higher bacterial diversity (~181 OTUs) than type I mats (Reid et al. 2000; Myshrall et al. 2010; Khodadad and Foster 2012; Mobberley et al. 2013).

Previous bacterial diversity studies based on 16S rDNA amplicons of Highborne Cay microbialites reveal that bacteria assigned to the Proteobacteria dominate all three mat types (52.0%), followed by Planctomycetes (12.1%), Cyanobacteria (11.6%), and Bacteroidetes (10.4%) (Myshrall et al. 2010; Mobberley et al. 2013). Within the Proteobacteria, Alphaproteobacteria are dominant, making up to 42% of the proteobacterial sequences, followed by 4.9% for both the Delta- and Gamma-proteobacterial lineages (Myshrall et al. 2010; Mobberley et al. 2013). The surrounding water communities differ from those in the stromatolitic mats I–III, with more taxa assigned to the Alphaproteobacteria and Gammaproteobacteria in the water, and no Deltaproteobacteria detected (Myshrall et al. 2010; Mobberley et al. 2013). Like Shark Bay, Proteobacteria are the dominant group in Highborne Cay, with Alphaproteobacteria being the most abundant OTU in the pink thrombolitic mats (Myshrall et al. 2010; Mobberley et al. 2013).

Metagenomes for both the microbial fraction and the viral fraction have been published for Highborne Cay (Desnues et al. 2008). However, because the surrounding environment was not sampled, including the water, it is hard to know which taxa are specific to the microbialite. Also, multiple displacement amplification (MDA) was used to obtain DNA for sequencing, which is known to affect quantitative results due to amplification bias (Abulencia et al. 2006; Yilmaz et al. 2010; Abbai et al. 2012).

The Highborne Cay viral metagenome was dominated by ssDNA microphages, likely due to MDA bias toward circular templates, which enriches these viral genotypes and skews quantitative results (Kim and Bae 2011). Moreover, these phages are present in the surrounding seawater (Desnues et al. 2008). Nonetheless, some ssDNA viral genotypes in Highborne Cay were found to be highly specific, suggesting that microbialites act as islands of viral diversity (Desnues et al. 2008). Highborne viral metagenomes had contigs that were similar to those in the Sargasso Sea or Cuatro Ciénegas. No Caudovirales phage sequences were detected, which is unusual, and may be the results of bias from MDA (Desnues et al. 2008).

5.2.4.3 Cuatro Ciénegas Microbialites: Pozas Azules II and Rios Mesquites

Cuatro Ciénegas, or the "four marshes," is found in the northern Mexican state of Coahuila. Cuatro Ciénegas is home to Pozas Azules II or the "blue pools," which have thrombolites, and Rios Mesquites ("mesquite tree rivers"), which have onco-lites (Breitbart et al. 2009). Cuatro Ciénegas has metagenomic data sets for both the viral and microbial fractions, but not samples from the sediment or water (Desnues et al. 2008; Breitbart et al. 2009).

Comparing the viral metagenomic data in Highborne Cay and Cuarto Ciénegas suggests that microbialites act as islands of novel viral genotypes and reflect an ancient marine origin that has persisted in a freshwater ecosystem (Desnues et al. 2008). However, high abundances of ssDNA viruses within both systems are an artifact of the bias of MDA toward circular templates, which enriches for ssDNA sequences and skews quantitative results (Abulencia et al. 2006; Yilmaz et al. 2010; Abbai et al. 2012). Pozas Azules II is similar to Highborne Cay, but it has more dsDNA phage sequences, including sequences that are similar to common marine phages, which could be due to cross-contamination from marine samples.

5.2.4.4   Pavilion Lake

Pavilion Lake (50.8°N, 121.7°W) lies in Mable Canyon between Squamish and Lillooet, BC, and contains active thrombolite-forming microbialites (Laval et al. 2000). Pavilion Lake thrombolitic microbialites range from ~2000 to 12,000 years in age, with the corrected age maximum at ~4600 years and the uncorrected age maximum at ~12,300, based on Uranium-Thorium dating (Laval et al. 2000). These age estimates put the initial formation of thrombolites in Pavilion Lake after the end of the last ice age (Laval et al. 2000).

Pavilion Lake is a dimictic, alkaline (pH 8.4), oligotrophic (mean total phosphorus 3.3 μg L^{-1}), freshwater lake with hard water (mean calcium carbonate 182 mg L^{-1}) (Laval et al. 2000; Lim et al. 2009; Brady et al. 2014; Omelon et al. 2013). Pavilion Lake microbialites consist primarily of calcite thrombolites, which change in morphotype as a function of depth, as follows: shallow domes resembling "open lettuce-beds" at ~10 m, intermediate domes resembling "cabbage heads" at ~20 m, intermediate-deep domes with hollow conical "asparagus-like" outcroppings at 25 m, and deep domes that possess "artichoke leaflets" at ≥25 m (Laval et al. 2000; Brady et al. 2014; Omelon et al. 2013; Fig. 5.5).

Morphological characteristics of the cyanobacterial community in Pavilion Lake microbialites have been described (Laval et al. 2000). The microbialites have thin (>5 mm) cyanobacterial-derived mats, comprising coccoids (*Gloeocapsa* spp.), rods (*Synechococcus* spp.), and filamentous cyanobacteria (*Fischerella* spp. and *Pseudoanabaena* spp.) (Laval et al. 2000). The water column of Pavilion Lake is dominated by *Synechococcus* spp. (8.3 × 10^4 cells per ml), the picoplankton known for causing carbonate precipitation "whiting events" (Thompson et al. 1997).

Pavilion Lake microbialites have round filamentous balls of cyanobacteria that are >2 mm at 15–20 m, which decrease in size with depth (Fig. 5.6, Brady et al. 2010). These filamentous balls of cyanobacteria are called "nodules," which range in pigmentation from purple to green to black. The purple cyanobacterial nodules were morphologically classified as either *Oscillatoria* sp. or *Calothrix* sp. (Laval et al. 2000). Recent clone library analysis suggests that the green nodules are *Leptolyngbya* sp. and the purple nodules are *Tolypothrix* sp. (Russell et al. 2014). *Tolypothrix* spp. and *Calothrix* spp. are morphologically similar, and their phylogenetic relationships are not well understood (Berrendero et al. 2011). Green and purple nodules create a microenvironment that has a higher pH and higher dissolved oxygen and has active lithification of calcite within the filaments (Brady et al. 2009, 2010). The nodules are

Fig. 5.5 Pavilion microbialite morphologies. Red scale bar is ~10 cm across for both. (Photo courtesy of Donnie Reid (Nuytco research) with permission)

Fig. 5.6 Pavilion Lake nodular cyanobacterial morphologies. Scale of the photo on the right at ~10 cm across. (Photo courtesy of Tyler Mckay (UC Davis))

only found at 15–25 m depths, which makes it unlikely that they are major players in the formation of the overall microbialite structure; otherwise they would be found at all depths.

The diversity and abundance of microbial taxa and the metabolic potential of Pavilion Lake microbial communities are unknown. Only a single metagenomic study is available for freshwater microbialites, which did not account for the surrounding environment (e.g., sediments and water column) (Breitbart et al. 2009). Is the microbialite microbial community and metabolic potential the same or different from that in the surrounding environment? This knowledge will either support previous studies that microbial community and metabolic potential are microbialite specific or will provide evidence of microbes from the surrounding environment (Breitbart et al. 2009; Khodadad and Foster 2012; Mobberley et al. 2013). A metagenomic study from Pavilion Lake suggests the metabolism, viral defense genes, and microbial abundances differ between surrounding water column, sediments, and microbialites (White III et al. 2016).

5.2.4.5   Kelly Lake

Kelly Lake (51.1°N 101.5°W) is located outside Clinton, BC, also within Mable Canyon, where Pavilion Lake is also located. Whispering Pines tribe among the Secwepemc (Shuswap) first nation having settled within Clinton, BC, since the last ice age nearly 10,000 years ago. Very little oral tradition exists from the Whispering

Pines tribe relationship with Kelly Lake to outside their community. Kelly Lake was named after Edward Kelly, and irishman in 1863 after a large ranch near the lake (Akrigg and Akrigg 1997)

Kelly Lake is ~40 km from Pavilion Lake, at an elevation of 1.07 km (Ferris et al. 1997). Kelly Lake is a smaller (1.5 × 0.3 km), is more shallow (maximum depth 35 m), and has lower water column visibility (<10 m Secchi disk depth) than Pavilion Lake (Ferris et al. 1997). The microbialites in Kelly Lake are mainly thrombolites, but stromatolites are also found and have morphologies similar to Pavilion Lake thrombolites (Ferris et al. 1997; Fig. 5.3). Kelly Lake has lower hardness of water (mean $CaCO_3$ ~163 mg L^{-1}) than Pavilion Lake, is less oligo-trophic than Pavilion Lake (mean phosphorus ~5.3 µg L^{-1}), and has half the salt (Na^+/K^+), total nitrogen, sulfate, and sulfide than Pavilion Lake (Ferris et al. 1997; Lim et al. 2009).

Cyanobacterial composition within Kelly Lake was been morphologically char-acterized *Synechococcus* sp., in the water column are ~105 ml^{-1} in Kelly Lake (Ferris et al. 1997). Morphologically, cyanobacteria in Kelly Lake are classified as coccoids (*Gloeocapsa* sp.), rods (*Synechococcus* sp.), and filaments (*Fischerella* sp. and *Pseudoanabaena* sp.), similar to those in Pavilion Lake (Ferris et al. 1997). Ferris et al. suggest that nodules and mats were present; however, current samples (2011) had no visible mat but rather a deep black layer on most of the microbialites (1997). Mat thickness was not measured in the study by Ferris et al., so it is unknown whether they had similar mat thickness or mats at all (1997). However, due to Kelly Lake proximity to Pavilion Lake, it is suggested that cyanobacterial mats of Kelly Lake microbialites at least at some point were similar to those of Pavilion Lake. The lake has changed since the 1997, and the high amounts of invertebrates on Kelly Lake microbialites were alarming in our 2011 sampling trip. These high amounts of metazoans indicate that their grazing could have reduced cyanobacterial mat in Kelly Lake. Modern Kelly Lake microbialites (at ~20 m) have deeply encrusted coccoid cyanobacteria (potentially *Gloeobacter* sp.) underneath the covering black layer of iron sulfide. Small green nodules were found on microbialites tree near the surface, but these were quite rare. Kelly Lake appears to be very different than Pavilion Lake. At present, no molecular survey exists for Kelly Lake.

5.2.4.6   Clinton Creek

Clinton Creek has a subarctic climate and is located near the Arctic Circle, ~77 km northwest of Dawson City, Yukon, Canada (64°26′42″N and 140°43′25″W). It is a flooded open pit from an abandoned asbestos mine, which was active from 1969 to 1978 (Power et al. 2011a). Clinton Creek has served as a model research site for carbon sequestration studies within industrial mining sites (Power et al. 2011a). Sediment mineralogy is highly variable, containing quartz, muscovite, kaolinite, and chrysotile, as well as minor aragonite and trace calcite (Power et al. 2011a).

Aragonite will preferentially lithify abiotically over calcite due to the inhibiting effect of the high magnesium to calcium ratio (Power et al. 2011a). Clinton Creek microbialites have a columnar thrombolitic morphology that is primarily composed of aragonite with spherulitic texture (Power et al. 2011a). The water in the open pit is subsaline (>3.0 g L^{-1}), alkaline (average pH 8.4), and supersaturated with aragonite (Power et al. 2011a). It is oligotrophic (undetectable phosphorus) and has low iron (~0.04 mg L^{-1}), and light penetrates to the bottom (Power et al. 2011a). The water is very hard (>200 mg L^{-1} and >75 mg Ca L^{-1}) and has high alkalinity (>190 mg HCO_3^- L^{-1}) (Power et al. 2011a). There is minimal nutrient input into the open pit due to the lack of surrounding soil and human activity over the last 35 years (Power et al. 2011a).

Power et al. (2011a) were the first to microscopically and isotopically characterize the microbial community of the Clinton Creek microbialite, which was morphologically described to have benthic diatoms (e.g., *Brachysira* spp.), filamentous algae (e.g., *Oedogonium* spp.), dinoflagellates, and cyanobacteria. Power et al. (2011a) noted the low microbial growth rates and suggested that the low sediment input had little influence geologically, but they did not examine microbe–sediment interactions.

The microbial community structure and metabolic potential of the Clinton Creek microbialites and their surrounding environment are uncharacterized, as are microbialite systems from cold regions in general. As well, the metabolic potential of freshwater microbialites, especially from cold environments, is mostly uncharacterized; only one freshwater microbialite metagenomic dataset exists, which is from a tropical environment and which was biased by using MDA for DNA amplification (Breitbart et al. 2009). Prior metagenomic studies, whether marine or freshwater, did not examine the composition or metabolic potential of the adjacent sediments or surrounding waters; such sampling would provide a context for identifying the microbes and processes that are confined to the microbialites. In Clinton Creek, the microbialite and sediment communities that were compared using metagenomics determined that sediment and microbialite metabolic potential and taxonomic abundances differed as well the microbialite metagenomes were more similar to polar microbial mats than over modern stromatolites from marine systems (White III et al. 2015). Photosynthetic was enriched in microbialites over sediments as well in Clinton Creek due to higher cyanobacterial abundances (White III et al. 2015).

5.3 Conclusions

Microbialites represent modern laboratories to test hypotheses related to ancient ecosystems that have global distributions. We can elucidate key questions using microbialites, such as "how does life assemble" and "how do communities of microbes form." Microbialites represent the past, present, and future of life on this planet. Finally, with modern techniques, we can learn from their history.

References

Abbai NS, Govender A, Shaik R, Pillay B (2012) Pyrosequence analysis of unamplified and whole genome amplified DNA from hydrocarbon-contaminated groundwater. Mol Biotechnol 50:39–48

Abulencia CB, Wyborski DL, Garcia JA, Podar M, Chen W, Chang SH (2006) Environmental whole-genome amplification to access microbial populations in contaminated sediments. Appl Environ Microbiol 72:3291–3330

Aitken JD (1967) Classification and environmental significance of cryptalgal limestones and dolomites, with illustrations from the Cambrian and Ordovician of southwestern Alberta. J Sediment Petrol 37:1163–1178

Akrigg V, Akrigg H (1997) British Columbia Place Names: Third Edition by G.P. (Philip). https://books.google.com/books?id=AVQ5RZeAFCkC&pg=PA133&lpg=PA133&dq=Kelly+lake+BC+named+after+an+irish+trapper&source=bl&ots=RvNLGI5_rn&sig=ACfU3U1wvLwv8MHZ_PjKEkVh2jbbz_kr0g&hl=en&sa=X&ved=2ahUKEwjez4elpYDpAhVIV80KHeveDlEQ6AEwA3oECAoQAQ#v=onepage&q=Kelly%20lake%20BC%20named%20after%20an%20irish%20trapper&f=false

Allen MA, Goh F, Burns BP, Neilan BA (2009) Bacterial, archaeal and eukaryotic diversity of smooth and pustular microbial mat communities in the hypersaline lagoon of Shark Bay. Geobiology 7:82–96

Allwood AC, Walter MR, Kamber BS, Marshall CP, Burch IW (2006) Stromatolite reef from the early Archaean era of Australia. Nature 441:714–718

Andersen DT, Sumner DY, Hawes I, Webster-Brown J, McKay CP (2011) Discovery of large conical stromatolites in Lake Untersee. Antarct Geobiol 9:280–293

Arp G, Thiel V, Reimer A, Michaelis W, Reitner J (1999) Biofilm exopolymers control microbialite formation at thermal springs discharging into the alkaline Pyramid Lake, Nevada, USA. Sediment Geol 126:159–176

Arp G, Reimer A, Reitner J (2004) Microbialite formation in seawater of increased alkalinity, Satonda Crater Lake, Indonesia. J Sediment Res 74:318–325

Awramik SM (1971) Precambrian columnar stromatolite diversity: reflection of metazoan appearance. Science 174:825–827

Babilonia J, Conesa A, Casaburi G, Pereira C, Louyakis AS, Reid RP, Foster JS (2018) Comparative metagenomics provides insight into the ecosystem functioning of the Shark Bay Stromatolites, Western Australia. Front Microbiol 9:1359

Barabesi CA, Galizzi G, Mastromei M, Rossi E, Tamburini B, Perito B (2007) *Bacillus subtilis* gene cluster involved in calcium carbonate biomineralization. J Bacteriol 189:228–235

Baumgartner LK, Reid RP, Dupraz C, Decho AW, Buckley DH, Spear JR, Przekop KM, Visscher PT (2006) Sulfate reducing bacteria in microbial mats: changing paradigms, new discoveries. Sediment Geol 185:131–145

Baumgartner LK, Spear JR, Buckley DH, Pace NR, Reid RP, Dupraz C, Visscher PT (2009) Microbial diversity in modern marine stromatolites, Highborne Cay, Bahamas. Environ Microbiol 11:2710–2719

Bernhard JM, Edgcomb VP, Visscher PT, McIntyre-Wressnig A, Summons RE, Bouxsein ML, Louis L, Jeglinski M (2013) Insights into foraminiferal influences on microfabrics of microbialites at Highborne Cay, Bahamas. Proc Natl Acad Sci 110:9830–9834

Berrendero E, Perona E, Mateo P (2011) Phenotypic variability and phylogenetic relationships of the genera *Tolypothrix* and *Calothrix* (Nostocales, Cyanobacteria) from running water. Int J Syst Evol Microbiol 61:3039–3051

Bischoff et al (2004) The springs of Lake Pátzcuaro: Chemistry, salt-balance, and implications for the water balance of the lake. Appl Geochem 19(11):1827–1835. https://doi.org/10.1016/j.apgeochem.2004.04.003

Bontognali TR, Sessions AL, Allwood AC, Fischer WW, Grotzinger JP, Summons RE, Eiler JM (2012) Sulfur isotopes of organic matter preserved in 345-billion-year-old stromatolites reveal microbial metabolism. Proc Natl Acad Sci 109:15146–15151

Boquet E, Boronat A, Ramos-Cormenzana A (1973) Production of calcite (calcium carbonate) crystals by soil bacteria is a general phenomenon. Nature 246:527–529

Bosak T, Greene SE, Newman DK (2007) A likely role for anoxygenic photosynthetic microbes in the formation of ancient stromatolites. Proc Natl Acad Sci 5:119–126

Bosak T, Liang B, Wu TD, Templer SP, Evans A, Vali H, Guerquin-Kern JL, Klepac-Ceraj V, Sim MS, Mui J (2012) Cyanobacterial diversity and activity in modern conical microbialites. Geobiology 10:384–401

Bottos EM, Vincent WF, Greer CW, Whyte LG (2008) Prokaryotic diversity of arctic ice shelf microbial mats. Environ Microbiol 10:950–966

Brady AL, Slater G, Laval B, Lim DS (2009) Constraining carbon sources and growth rates of freshwater microbialites in Pavilion Lake using ^{14}C analysis. Geobiology 7:544–555

Brady AL, Slater GF, Omelon CR, Southam G, Druschel G, Andersen DT, Hawes I, Laval B, Lim DSS (2010) Photosynthetic isotope biosignatures in laminated micro-stromatolitic and non-laminated nodules associated with modern, freshwater microbialites in Pavilion Lake, BC. Chem Geology 274:56–67

Brady AL, Laval B, Lim DSS, Slater GF (2014) Autotrophic and heterotrophic associated biosignatures in modern freshwater microbialites over seasonal and spatial gradients. Org Geochem 67:8–18

Braithwaite CJR, Zedefa V (1994) Living hydromagnesite stromatolites from Turkey. Sediment Geol 92:1–5

Breitbart M, Hoare A, Nitti A, Siefert J, Haynes M, Dinsdale E, Edwards R, Souza V, Rohwer F, Hollander D (2009) Metagenomic and stable isotopic analyses of modern freshwater microbialites in Cuatro Ciénegas, Mexico. Environ Microbiol 11:16–34

Bridgwater DN, Holmes AJ, O'Hara LS (1999) Complex controls on the trace-element chemistry of non-marine ostracods: an example from Lake Patzcuaro, Central Mexico. Palaeogeogr Palaeoclimatol Palaeoecol 148:117–131

Buick R, Groves DI, Dunlop JS (1995) Abiologial origin of described stromatolites older than 3.2 Ga. Geology 23:191

Burne RV, Moore LS (1987) Microbialites: organosedimentary deposits of benthic microbial communities. PALAIOS 2:241–254

Burns BP, Goh F, Allen M, Neilan BA (2004) Microbial diversity of extant stromatolites in the hypersaline marine environment of Shark Bay, Australia. Environ Microbiol 6:1096–1101

Canfield DE, Abicht K, Thamdrup B (2000) The archaean sulfur cycle and the early history of atmospheric oxygen. Science 288:658–661

Canfield DE (2005) The early history of atmospheric oxygen: homage to Rober A. Garrels. Annu Rev Earth Planet Sci 33:1–36

Castanier S, Le Metayer-Levrel G, Perthuisot JP (1999) Ca-carbonates precipitation and limestone genesis the microbiogeologist point of view. Sediment Geol 126:9–23

Centeno CM, Legendre P, Beltrán Y, Alcántara-Hernández RJ, Lidström UE, Ashby MN, Falcón LI (2012) Microbialite genetic diversity and composition relate to environmental variables. FEMS Microbiol Eco. 82:724–735

Clokie M, Clokie JR, Millard DA, Mann HN (2010) T4 genes in the marine ecosystem: studies of the T4-like cyanophages and their role in marine ecology. Virol J 7:291

Couradeau E, Benzerara K, Moreira D, Gérard E, Kaźmierczak J, Tavera R, López-García P (2011) Prokaryotic and eukaryotic community structure in field and cultured microbialites from the alkaline Lake Alchichica (Mexico). PLoS One 6:e28767

Crowe SA, Døssing LN, Beukes NJ, Bau M, Kruger SJ, Frei R, Canfield DE (2013) Atmospheric oxygenation three billion years ago. Nature 501:535–538

Desnues C, Rodriguez-Brito B, Rayhawk S, Kelley S, Tran T, Haynes M, Liu H, Furlan M, Wegley L, Chau B, Ruan Y et al (2008) Biodiversity and biogeography of phages in modern stromatolites and thrombolites. Nature 452:340–343

Dhami NK, Reddy MS, Mukherjee A (2013) Biomineralization of calcium carbonate polymorphs by the bacterial strains isolated from calcareous sites. J Microbiol Biotechnol 23:707–714

Dhami NK, Reddy MS, Mukherjee A (2014) Synergistic role of bacterial urease and carbonic anhydrase in carbonate mineralization. Appl Biochem Biotechnol 172:2552–2561

Dick J, De Windt W, De Graef B, Saveyn H, Van der Meeren P, De Belie N, Verstraete W (2006) Bio-deposition of a calcium carbonate layer on degraded limestone by *Bacillus* species. Biodegradation 17:357–367

Dittricha M, Müllera B, Mavrocordatosb D, Wehrlia B (2003) Induced calcite precipitation by cyanobacterium *Synechococcus*. Acta Hydrochim Hydrobiol 31:162–169

Dupraz C, Visscher PT (2005) Microbial lithification in marine stromatolites and hypersaline mats. Trends Microbiol 13:429–438

Dupraz C, Reid RP, Braissant O, Decho AW, Norman RS, Visscher PT (2009) Processes of carbonate precipitation in modern microbial mats. Earth Sci Rev 96:141–162

Edgcomb VP, Bernhard JM, Summons RE, Orsi W, Beaudoin D, Visscher PT (2013) Active eukaryotes in microbialites from Highborne Cay, Bahamas, and Hamelin Pool (Shark Bay), Australia. ISME J 8(2):418

Elser JJ, Schampel JH, García-Pichel F, Wade BD, Souza V, Eguiarte L, Escalante A, Farmer JD (2005) Effects of phosphorous enrichment and grazing snails on modern stromatolitic microbial communities. Freshw Biol 50:1808–1825

Ercole C, Cacchio P, Botta AL, Centi V, Lepidi A (2007) Bacterially induced mineralization of calcium carbonate: the role of exopolysaccharides and capsular polysaccharides. Microsc Microanal 13:42–50

Farías ME, Rascovan N, Toneatti DM, Albarracın VH, Flores MR, Poire DG, Collavino MM, Aguilar OM, Vazquez MP, Polerecky L (2012) The discovery of stromatolites developing at 3570 m above sea level in a high-altitude volcanic lake Socompa, Argentinean Andes. PLoS One 8:e53497

Ferris FG, Thompson JB, Beveridge TJ (1997) Modern freshwater microbialites from Kelly Lake, British Columbia, Canada. PALAIOS 12:213–219

Friedrich CG, Bardischewsky F, Rother D, Quentmeier A, Fischer J (2005) Prokaryotic sulfur oxidation. Curr Opin Microbiol 8:253–259

Gallagher KL, Kading TJ, Braissant O, Dupraz C, Visscher PT (2012) Inside the alkalinity engine: the role of electron donors in the organomineralization potential of sulfate-reducing bacteria. Geobiology 10:518–530

Garrett P (1970) Phanerozoic stromatolites: noncompetitive ecologic restriction by grazing and burrowing animals. Science 169:171–173

Gischler E, Gibson MA, Oschmann W (2008) Giant Holocene freshwater microbialites, Laguna Bacalar, Quintana Roo, Mexico. Sedimentology 55:1293–1309

Goh F, Allen MA, Leuko S, Kawaguchi T, Decho AW, Burns BP, Neilan BA (2009) Determining the specific microbial populations and their spatial distribution within the stromatolite ecosystem of Shark Bay. ISME J 3:383–396

Graham LE, Knack JJ, Piotrowski MJ, Wilcox LE, Cook ME, Wellman CH, Taylor W, Lewis LA, Arancibia-Avila P (2014) Lacustrine *Nostoc* (Nostocales) and associated microbiome generate a new type of modern clotted microbialite. J Phycol 50:280–291

Grey K, Moore LS, Burne RV, Pierson BK, Bauld J (1990) Lake Thetis, Western Australia: an example of saline sedimentation dominated by benthic microbial processes: Aus. J Mar Freshw Res 41:275–300

Grotzinger JP, Rothman DR (1996) An abiotic model for stromatolite morphogenesis. Nature 383:423–425

Grotzinger JP, Knoll AH (1999) Stromatolites in Precambrian carbonates: evolutionary mileposts or environmental dipsticks? Annu Rev Earth Planet Sci 27:313–358

Hammes F, Boon N, de Villiers J, Verstraete W, Siciliano SD (2003) Strain-specific ureolytic microbial calcium carbonate precipitation. Appl Environ Microbiol 69:4901–4909

Havemann SA, Foster JS (2008) Comparative characterization of the microbial diversities of an artificial microbialite model and a natural stromatolite. Appl Environ Microbiol 74:7410–7421

Hawes I, Sumner DY, Andersen DT, Mackey TJ (2011) Legacies of recent environmental change in the benthic communities of Lake Joyce, a perennially ice-covered Antarctic lake. Geobiology 9:394–410

Jansson C, Northern T (2010) Calcifying cyanobacteria-the potential of biomineralization for carbon capture and storage. Curr Opin Biotechnol 21:365–371

Kalkowsky E (1908) Oolith und Stromatolith im norddeutschen Buntsandstein. Zeitschrift Deutschen geol. Gesellschaft 60: 68–125 (Article in German)

Kazmierczak J, Kempe S (2006) Genuine modern analogues of Precambrian stromatolites from caldera lakes of Niuafo'ou Island, Tonga. Naturwissenschaften 93:119–126

Kempe S, Kazmierczak J, Landmann G, Konuk T, Reimer A, Lipp A (1991) Largest known microbialites discovered in Lake Van, Turkey. Nature 349:605–608

Khodadad CL, Foster JS (2012) Metagenomic and metabolic profiling of nonlithifying and lithifying stromatolitic mats of Highborne Cay, The Bahamas. PLoS One 7:e38229

Kim KH, Bae JW (2011) Amplification methods bias metagenomic libraries of uncultured single-stranded and double-stranded DNA viruses. Appl Environ Microbiol 77:7663–7668

Klassen JL (2010) Phylogenetic and evolutionary patterns in microbial carotenoid biosynthesis are revealed by comparative genomics. PLoS One 5:e11257

Köcher S, Jurgen B, Müller V, Sandmann G (2009) Structure, function and biosynthesis of carotenoids in the moderately halophilic bacterium *Halobacillus halophilus*. Arch Microbiol 191:95–104

Konishi Y, Prince J, Knott B (2001) The fauna of thrombolitic microbialites, Lake Clifton, Western Australia. Hydrobiologia 457:39–47

Laval B, Cady SL, Pollack JC, McKay CP, Bird JS, Grotzinger JP, Ford DC, Bohm HR (2000) Modern freshwater microbialite analogues for ancient dendritic reef structures. Nature 407:626–629

Lee YN (2003) Calcite production by *Bacillus amyloliquefaciens* CMB01. J Microbiol 41:345–348

Lim DSS, Laval BE, Slater G, Antoniades D, Forrest AL, Pike W, Pieters R, Saffari M, Reid D, Schulze-Makuch D, Andersen D, McKay CP (2009) Limnology of Pavilion Lake B.C. - characterization of a microbialite-forming environment. Fund Appl Limnol 173:329–351

Lionard M, Péquin B, Lovejoy C, Vincent WF (2012) Benthic cyanobacterial mats in the high arctic: multi-layer structure and fluorescence responses to osmotic stress. Front Microbiol 3:140

Logan BW, Rezak R, Ginsburg RN (1964) Classification and environmental significance of algal stromatolites. J Geology 72:68–83

López-García P, Kazmierczak J, Benzerara K, Kempe S, Guyot F, Moreira D (2005) Bacterial diversity and carbonate precipitation in the giant microbialites from the highly alkaline Lake Van, Turkey. Extremophiles 9:263–274

Lowe DR (1994) Abiological origin of described stromatolites older than 3.2 Ga. Geology 22:387–390

Lundberg J, Mcfalane D (2011) Subaerial freshwater phosphatic stromatolites in Deer Cave, Sarawak a unique geobiological cave formation. Geomorphology 128:57–72

Lyons TW, Reinhard CT, Planavsky NJ (2014) The rise of oxygen in Earth's early ocean and atmosphere. Nature 506:307–315

Maliva GR, Missimer MT, Leo CK, Statom AR, Dupraz C, Lynn M, JAD D (2000) Unusual calcite stromatolites and pisoids from a landfill leachate collection system. Geology 28:931–934

McLoughlin N, Wilson LA, Brasier MD (2008) Growth of synthetic stromatolites and wrinkle structures in the absence of microbes - implications for the early fossil record. Geobiology 6:95–105

Mobberley JM, Khodadad CL, Foster JS (2013) Metabolic potential of lithifying cyanobacteria-dominated thrombolitic mats. Photosynth Res 118:125–140

Myshrall KL, Mobberley JM, Green SJ, Visscher PT, Havemann SA, Reid RP, Foster JS (2010) Biogeochemical cycling and microbial diversity in the thrombolitic microbialites of Highborne Cay, Bahamas. Geobiology 8:337–354

Neilan BA, Pearson LA, Muenchhoff J, Moffitt MC, Dittmann E (2013) Environmental conditions that influence toxin biosynthesis in cyanobacteria. Environ Microbiol 5:1239–1253

Nübel U, Garcia-Pichel F, Kühl M, Muyzer G (1999) Quantifying microbial diversity: morpho-types, 16S rRNA genes, and carotenoids of oxygenic phototrophs in microbial mats. Appl Environ Microbiol 65:422–430

Obst M, Wehrli B, Dittrich M (2009) CaCO3 nucleation by cyanobacteria: laboratory evidence for a passive, surface-induced mechanism. Geobiology 7:324–347

Omelon CR, Brady AL, Slater GF, Laval B, Lim DSS, Southam G (2013) Microstructure variability in freshwater microbialites, Pavilion Lake, Canada. Palaeogeogr Palaeoclimat 392:62–70

Paerl HW, Steppe TF, Reid RP (2001) Bacterially mediated precipitation in marine stromatolites. Environ Microbiol 3:123–130

Papineau D, Walker JJ, Mojzsis SJ, Pace NR (2005) Composition and structure of microbial communities from stromatolites of Hamelin Pool in Shark Bay, Western Australia. Appl Environ Microbiol 71:4822–4832

Patterson CC (1956) Age of meteorites and the Earth. Geochim Cosmochim Acta 10:230–237

Perry RS, Mcloughlin N, Lynne BY, Sephton MA, Oliver JD, Perry CC, Campbell K, Engel MH, Farme JD, Brasier MD, Staley JT (2007) Defining biominerals and organominerals: direct and indirect indicators of life. Sediment Geol 201:157–179

Pia J (1933) Die Rezenten Kalkesteine: Zeitschr. fiir Kristallographle, Mineralogie und Petrographic, Abt. B, 1-420 (In German)

Planavsky NJ, Asael D, Hofmann A, Reinhard CT, Lalonde SV, Knudsen A, Wang X, Ossa FO, Pecoits E et al (2014) Evidence for oxygenic photosynthesis half a billion years before the Great Oxidation Event. Nat Geosci 7:283–286

Power IM, Wilson SA, Thom JM, Dipple GM, Southam G (2007) Biologically induced mineralization of dypingite by cyanobacteria from an alkaline wetland near Atlin, British Columbia, Canada. Geochem Trans 8:13

Power IM, Wilson SA, Dipple GM, Southam G (2011a) Modern carbonate microbialites from an asbestos open pit pond, Yukon, Canada. Geobiology 9:180–195

Power IM, Wilson SA, Small DP, Dipple GM, Wan W, Southam G (2011b) Microbially mediated mineral carbonation: roles of phototrophy and heterotrophy. Environ Sci Technol 45:9061–9068

Rebollar EA, Avitia M, Eguiarte LE, González-González A, Mora L, Bonilla-Rosso G, Souza V (2012) Water-sediment niche differentiation in ancient marine lineages of *Exiguobacterium* endemic to the Cuatro Ciénegas Basin. Environ Microbiol 14:2323–2333

Reid RP, Visscher PT, Decho AW, Stolz JF, Bebout BM, Dupraz C, Macintyre IG, Paerl HW, Pinckney JL, Prufert-Bebout L, Steppe TF, DesMarais DJ (2000) The role of microbes in accretion, lamination and early lithification of modern marine stromatolites. Nature 406:989–999

Riding (2011) Microbialites, Stromatolites, and Thrombolites. https://doi.org/10.1007/978-1-4020-9212-1_196

Rothschild LJ, Mancinelli RL (1990) Model of carbon fixation in microbial mats from 3,500 Myr ago to the present. Nature 345:710–712

Russell JA, Brady AL, Cardman Z, Slater GF, Lim DS, Biddle JF (2014) Prokaryote populations of extant microbialites along a depth gradient in Pavilion Lake, British Columbia, Canada. Geobiology 12:250–264

Ruvindy R, White RA III, Neilan BA, Burns BP (2016) Unravelling core microbial metabolisms in the hypersaline microbialites of Shark Bay using high-throughput metagenomics. ISME J 10:183–196

Santos F, Peña A, Nogales B, Soria-Soria E, Del Cura MA, González-Martín JA, Antón J (2010) Bacterial diversity in dry modern freshwater stromatolites from Ruidera Pools Natural Park, Spain. Syst Appl Microbiol 33:209–221

Schneider D, Arp G, Reimer A, Reitner J, Daniel R (2013) Phylogenetic analysis of a microbialite-forming microbial mat from a hypersaline lake of the Kiritimati atoll, Central Pacific. PLoS One 8:e66662

Schopf JW (2006) Fossil evidence of Archean life. Philos Trans R Soc B 361:869–885

Smith AM, Andrews JE, Uken R, Thackeray Z, Perissinotto R, Leuci R, Marca-Bel A (2011) Rock pool tufa stromatolites on a modern South African wave-cut platform: partial analogues for Archaean stromatolites? Terra Nova 23:375–381

Spadafora A, Perri E, McKenzie JA, Vasconcelos C (2010) Microbial biomineralization processes forming modern Ca:Mg carbonate stromatolites. Sedimentology 57:27–40

Sprachta S, Camoin G, Golubic S, Le-Campion T (2001) Microbialites in a modern lagoonal environment: nature and distribution, Tikehau atoll (French Polynesia). Palaeogeo Palaeoclim Palaeoecol 175:103–124

Squyres SW, Andersen DW, Nedell SS, Wharton RA Jr (1991) Lake Hoare, Antarctica: sedimentation through a thick perennial ice cover. Sedimentology 38:363–379

Stal LJ (2000) Cyanobacterial mats and stromatolites. In: Whiton BA, Potts M (eds) Ecology of cyanobacteria: their diversity in time and space. Kluwer Academic Publications, Dordrecht, pp 61–120

Suosaari EP, Reid RP, Playford PE, Foster JS, Stolz J, Casaburi G, Hagan PD, Chirayath V, Macintyre IG, Planavsky NJ, Eberli GP (2016) New multi-scale perspectives on the stromatolites of Shark Bay, Western Australia. Sci Rep 6:20557

Suttle AS (2005) Viruses in the sea. Nature 437:356–336

Tarhan LG, Planavsky NJ, Laumer CE, Stolz JF, Reid RP (2013) Microbial mat controls on infaunal abundance and diversity in modern marine microbialites. Geobiology 11:485–497

Thompson JB, Ferris GF (1990) Cyanobacterial precipitation of gypsum, calcite, and magnesite from natural alkaline lake water. Geology 18:995–998

Thompson JB, Schultze-Lam S, Beveridge TJ, Des Marais DJ (1997) Whiting events: biogenic origin due to the photosynthetic activity of cyanobacteria picoplankton. Limnol Oceanogr 42:133–141

Thompson RL, Zeng Q, Kelly L, Huang HK, Singer UA, Stubbea J, Chisholm WS (2011) Phage auxiliary metabolic genes and the redirection of cyanobacterial host carbon metabolism. Proc Natl Acad Sci 108:E757–E764

Van Kranendonk MJ, Philippot P, Lepot K, Bodorkos S, Pirajno F (2008) Geological setting of Earth's oldest fossils in the ca. 3.5 Ga Dresser Formation, Pilbara Craton, Western Australia. Precambrian Res 167:93–124

Varin T, Lovejoy C, Jungblut AD, Vincent WF, Corbeil J (2012) Metagenomic analysis of stress genes in microbial mat communities from Antarctica and the High Arctic. Appl Environ Microbiol 78:549–559

Vasconcelos C, Warthmann R, McKenzie JA, Visscher PT, Bittermann AG, van Lith Y (2006) Lithifying microbial mats in Lagoa Vermelha, Brazil: modern Precambrian relics? Sediment Geol 185:175–183

Visscher PT, Reid RP, Bebout BM (2000) Microscale observations of sulfate reduction: correlation of microbial activity with lithified micritic laminae in modern marine stromatolites. Geology 28:919–922

Wacey D, Kilburn RM, Saunders M, Cliff J, Brasier DM (2011) Microfossils of sulphur-metabolizing cells in 3.4-billion-year-old rocks of Western Australia. Nat Geosci 4:698–702

Warden JG, Casaburi G, Omelon CR, Bennett PC, Breecker DO, Foster JS (2016) Characterization of microbial mat microbiomes in the modern thrombolite ecosystem of Lake Clifton, Western Australia using shotgun metagenomics. Front Microbiol 7:1064

White RA III, Grassa CJ, Suttle CA (2013a) First draft genome sequence from a member of the genus *Agrococcus*, isolated from modern microbialites. Genome Announc 1:e00391–e00313

White RA III, Grassa CJ, Suttle CA (2013b) Draft genome sequence of *Exiguobacterium pavilionensis* strain RW-2, with wide thermal, salinity, and pH tolerance, isolated from modern freshwater microbialites. Genome Announc 1:e00597–e00513

White RA III, Suttle CA (2013) The Draft Genome Sequence of *Sphingomonas paucimobilis* strain HER1398 (Proteobacteria), host to the giant PAU phage, indicates that it is a member of the genus *Sphingobacterium* (Bacteroidetes). Genome Announc 1:e00598–e00513

White RA III, Power IM, Dipple GM, Southam G, Suttle CA (2015) Metagenomic analysis reveals that modern microbialites and polar microbial mats have similar taxonomic and functional potential. Front Microbiol 6:966

White RA III, Chan AM, Gavelis GS, Leander BS, Brady AL, Slater GF et al (2016) Metagenomic analysis suggests modern freshwater microbialites harbour a distinct core microbial community. Front Microbiol 6:1531

White RA III, Wong HL, Ruvindy R, Neilan BA, Burns BP (2018) Viral communities of Shark Bay modern stromatolites. Front Microbiol 9:1–15

White RA III, Gavelis G, Soles SA, Gosselin E, Slater GF, Lim DSS, ... (2018) The complete genome and physiological analysis of the microbialite-dwelling Agrococcus pavilionensis sp. nov; reveals genetic promiscuity and predicted adaptations to environmental stress. Front Microbiol. https://www.frontiersin.org/articles/10.3389/fmicb.2018.02180/full

White RA III, Soles SA, Gavelis G, Gosselin E, Slater GF, Lim DSS, .. (2019) The complete genome and physiological analysis of the eurythermal firmicute exiguobacterium chiriqhucha strain RW2 isolated from a freshwater microbialite, widely adaptable to broad thermal, pH, and salinity ranges. Front Microbiol. https://www.frontiersin.org/articles/10.3389/fmicb.2018.03189/full

Wong HL, Smith DL, Visscher PT, Burns BP (2015) Niche differentiation of bacterial communities at a millimeter scale in Shark Bay microbial mats. Sci Rep 5:15607

Wong HL, Visscher PT, White RA III, Smith DL, Patterson M, Burns BP (2017) Dynamics of archaea at fine spatial scales in Shark Bay mat microbiomes. Sci Rep 7:46160

Wong HL, White RA, Visscher PT, Charlesworth JC, Vázquez-Campos X, Burns BP (2012) Disentangling the drivers of functional complexity at the metagenomic level in Shark Bay microbial mat microbiomes. ISME J 12(11):2619–2639

Wu YS, Yu GL, Li RH, Song LR, Jiang HX, Riding R, Liu LJ, Liu DY, Zhao R (2014) Cyanobacterial fossils from 252 Ma old microbialites and their environmental significance. Sci Rep 4:3820

Yilmaz S, Allgaier M, Hugenholtz P (2010) Multiple displacement amplification compromises quantitative analysis of metagenomes. Nat Methods 7:943–944

Zhou J, He Q, Hemme CL, Mukhopadhyay A, Hillesland K, Zhou A, He Z, Van Nostrand JD, Hazen TC, Stahl DA, Wall JD, Arkin AP (2011) How sulphate-reducing microorganisms cope with stress: lessons from systems biology. Nat Rev Microbiol 9:452–466

Chapter 6
The Importance of the Rare Biosphere for Astrobiological Studies and the Diversification and Resilience of Life on Earth

Jazmín Sánchez-Pérez, Juan Diego Villar, Nohely Alvarez-López,
Bernardo Águila, Jhoselinne Buenrostro, Luis J. Chino-Palomo,
Marisol Navarro-Miranda, Julián Felipe Cifuentes, Ana G. Cruz-Cruz,
Benjamín Vega-Baray, Mariette Viladomat, Maria Kalambokidis,
Luis E. Eguiarte, and Valeria Souza

Abstract Let's travel in our mind eye to either early Earth or Mars: life had already evolved, but it was most probably patchy and hard to find; it was likely rare. We can even imagine the same process in exoplanets or moons, life starting as an organic oddity that eventually starts to evolve by Darwinian evolution, consuming resources to grow and reproduce. Under that scenario, we can speculate that after the last universal common ancestor, LUCA, diversified locally, life kept on diversifying in

J. Sánchez-Pérez · B. Águila · M. Navarro-Miranda · M. Viladomat · M. Kalambokidis ·
L. E. Eguiarte · V. Souza (✉)
Instituto de Ecología, Universidad Nacional Autónoma de México, UNAM,
Mexico City, Mexico
e-mail: souza@unam.mx

J. D. Villar · J. F. Cifuentes
Pontificia Universidad Javeriana, Bogotá, DC, Colombia

N. Alvarez-López
Escuela Nacional de Estudios Superiores, Unidad Morelia, Universidad Nacional Autónoma
de México, UNAM, Morelia, Michoacán, Mexico

J. Buenrostro · A. G. Cruz-Cruz
Facultad de Estudios Superiores Iztacala, Universidad Nacional Autónoma de México,
UNAM, Tlalnepantla, Estado de México, Mexico

L. J. Chino-Palomo
Facultad de Medicina, Universidad Nacional Autónoma de México, UNAM,
Mexico City, Mexico

B. Vega-Baray
Instituto de Investigaciones Biomédicas, Universidad Nacional Autónoma de México,
UNAM, Mexico City, Mexico

© Springer Nature Switzerland AG 2020   135
V. Souza et al. (eds.), *Astrobiology and Cuatro Ciénegas Basin as an Analog of
Early Earth*, Cuatro Ciénegas Basin: An Endangered Hyperdiverse Oasis,
https://doi.org/10.1007/978-3-030-46087-7_6

a multitude of non-abundant lineages that were locally adapted and perhaps may explain why we see such early radiation in the tree of life. Therefore, life started as a rare event and the rare biosphere, that represent most of the taxonomic and functional diversity in actual microbial communities, probably represents the way communities have organized themselves since the beginning of the conquest of the Earth. In this chapter, we will explain what it means to have a low abundance in a community, how we study rarity, and why it is relevant for astrobiology.

6.1 What Does It Means to Be Rare and Its Importance in Microbial Communities?

For most of the last century, the understanding of microbial diversity was still based solely on the identification of phenotypic and morphometric characteristics through the use of microscopy by culture-dependent experimentation (Lynch and Neufeld 2015). In recent years, sequencing of the 16S rRNA gene has been broadly used for the study of microbial communities. The first extensive work that incorporated these new methods took place in the North Atlantic Ocean in the early 2000s (Sogin et al. 2006). This study revealed that most microbial communities had a heterogeneous and asymmetric distribution of species, where relatively few dominant species coexist with a high number of low-abundant species (Sogin et al. 2006; Nemergut et al. 2011). It was therefore believed that niches with a wide range of microorganisms at low frequencies were rare, due to the low performance of these species in different environments. Today, nearly 15 years later, shotgun metagenomics has confirmed that in most communities, the existence of the rare biosphere is a general feature of microbial communities. This technique not only seeks to characterize species, but it also estimates the magnitude of the functional potential of the community, in particular showing us the potential genomes of non-cultivable and undescribed microorganisms (Scholz et al. 2012).

Due to the initial low coverage of the sample, in the first studies of microbial communities, the phylotypes with extremely small populations were found only in specific niches, such as in marine sponge parenchyma (Taylor et al. 2012), or in amphibian skin (Walke et al. 2014), among others. One group (Gobet et al. 2012) identified low-abundance bacterial taxa in coastal sands in response to nutrient stress. Likewise, another group (Bowen et al. 2009) found a large composition of low-abundance species within marsh sediments, which were highly distinct from samples of the nearby water column. As a result, it was erroneously assumed that these species were not well adapted and therefore not important for the function and structure of these communities. However, through further studies, this vision changed.

First, let's specify that there are different definitions and approaches to measuring rarity. Though rarity concepts were initially applied to plants and animals, they are nonetheless easily extrapolated to microorganisms. The first type of rarity is defined by an organism with low abundance, which describes species that are found at a low frequency despite the size of the distribution area. Similarly, species can also be considered rare if it is only found in very specific niches; many of the existing examples include extremophilic microorganisms. Finally, there are organisms that are delimited by the geographical area in which they live, being almost endemic to a site.

The first kind of rarity we will explore is the low abundance. There are different ways in which a species can become low in abundance, such as through stochastic mechanisms that cause an abrupt change in the population (e.g., being uncompetitive in a new environment) or the result of a recent migration. In contrast, low abundance can also be a life strategy for many species either because that way they escape predators or because they use low abundance resources that may be provided by the rest of the community. For this reason, having low abundance allows for the survival of organisms in a trade-off mechanism, and it is very possible that rarity confers an adaptive advantage. This effect has been seen at the genotypic level with a phenomenon called frequency-dependent negative selection, where a genotype fitness is inversely proportional to its frequency in the population. This mechanism causes a genotype to initially increase its frequency when they are rare and reduce it when they become common. This process of succession of genotypes prevents one genotype from becoming fixed, thus maintaining diversity at different levels, as seen in mitochondria (Kazancıoğlu and Arnqvist 2014), clonal populations (Weeks and Hoffmann 2008), and among different cell strains (Minter et al. 2015). Moreover, predator and parasite dynamics can also modify the frequency of a bacterial population (Dybdahl and Lively 1998), where rare organisms are less predated compared to common ones (Winter et al. 2010).

The large proportion of rare species found in diverse ecosystems (Galand et al. 2009; Sogin et al. 2006; Lynch and Neufeld 2015) has led to further study the role of these organisms in microbial communities, as well as their ecological relevance. This rare biosphere, defined as all the organisms with a low frequency, is a relevant aspect of the microbial communities, as it includes the greatest amount of diversity within the community (Locey and Lennon 2016; Elshahed et al. 2008) and, in some cases, the largest proportion of the community (Sogin et al. 2006; Lynch and Neufeld 2015). Although, it is possible that the rarity of some species is a byproduct of processes like local extinction or low competitiveness in the community, many rare taxa could remain rare in the community over long periods (De Anda et al. 2018). This constancy indicates that they may play an important role in the community. If these organisms are not a stochastic product of migration or extinction and are preserved in microbial communities, then: What role do they play within the community?

One of the possible roles of the rare biosphere may be the contribution of diversity contained as a stability factor. Diversity is associated with the resilience of a system and functional capacity; so species perform functions in a given environmental condition. This aspect may be counterintuitive, given the high functional redundancy expected (Rousk et al. 2009). With so many organisms sharing functions, the rare biosphere does not seem necessary to maintain them; however, species with functions considered irrelevant could become important in another environmental condition, either by adding a new trait to the community or participating in new interspecific interactions (Fetzer et al. 2015). The rare biosphere could, therefore, function as a seed bank. By preserving diversity, different metabolic functions can be conserved over time, becoming beneficial when different environmental conditions arise. This is why the community, through its seed bank, can react and effectively cope with different environmental fluctuations.

Furthermore, rare organisms seem to have a different role in the maintenance of communities compared to abundant species, since they react to different stress factors (Zhang et al. 2018). For example, one of the most important roles reported is the response to pollution by degradation of organic compounds (Zhang et al. 2018; Aanderud et al. 2015; Wang et al. 2017). The degradation of organic compounds is carried out through a network of complex metabolic pathways that are shared and coupled along with several taxa (Fuentes et al. 2014). Many of these functions are associated with the rare biosphere since experimental studies have reported that the capacity to degrade pollutants and toxins is reduced when rare microorganisms are removed from wastewater (Hernandez-Raquet et al. 2013).

Additionaly, rare biosphere, with its large repertoire of functions, seems to be part of the complex metabolic pathways associated with biogeochemical cycles. By providing new substrates, these cycles play a key role in the assembly of communities. Low-abundant species of green and purple sulfur bacteria appear to be crucial for the absorption of carbon and nitrogen in aquatic communities (Musat et al. 2008). Likewise, there are several examples associated with different nutrient cycles, such as the use of alternative carbon sources as a nutritional substrate (Hernandez-Raquet et al. 2013; Mallon et al. 2015; Matos et al. 2005; Franklin et al. 2001) and in the different phases of the nitrogen cycle, specifically in nitrification (Griffiths et al. 2004), nitrogen uptake (Musat et al. 2008), and denitrification (Philippot et al. 2013).

Another role of the rare biosphere in a micorbial community is against invasive species (Vivant et al. 2013). As a large part of the diversity, low-abundant organisms manage to occupy many niches within the community. Therefore, the limited availability of niches and resources succeeds in suppressing the invasion of new organisms into the community (van Elsas et al. 2012). The establishment of an invasive species is dependent on the amount of resources not consumed by local species, as well as the rate at which both consume the existing nutrients: in other words, who is consuming nutrients faster (Tilman 1999). Thus, when a niche is not occupied by the native species, the invasion by an exogenous species can be favored (Pedrós-Alió 2012). In this way, the greater the diversity and the greater the number of organisms that consume different things, the less likely it is that there will be excess resources to enable the entry and establishment of an invasive species.

6.2 Dynamism in the Community: Succession Between Common and Rare

Currently, metagenomic analyses have shown the presence of thousands of new species in all the explored ecosystems, which are in low abundance in the community (Pedrós-Alió 2012). This ability to describe the different species that inhabit a particular community made it possible to compare their presence and abundance in different locations. On some occasions, members of the abundant species in one community may be part of the rare biosphere in another. In this case, certain species may be abundant in a particular environment, but when dispersed to other communities, they become rare (Lynch and Neufeld 2015). The succession between abundance and rarity can then be determined by environmental conditions and biotic interactions.

The rare biosphere can respond to environmental fluctuations in two ways: (1) the "conditionally rare", whose frequency is determined by certain environmental factors, so when there is a favorable change in environmental conditions, its population frequency increases (Coveley et al. 2015; Sogin et al. 2006); and (2) the "functionally rare", who after a favorable environmental fluctuation, does not increase their growth rate but rather increase their transcriptional activity (Hausmann et al. 2019). Whatever their response is (increase in number or increase in activity), this low-abundant genetic diversity becomes crucial to maintaining ecosystem functions during stressful conditions (Shade et al. 2012). Therefore, understanding the process of succession and exchange between rare and common species would provide a great deal of information about the dynamics of the communities, including how they are structured, their response to environmental fluctuations, and the resilience of the community provided by the seed bank of low-abundance organisms that are preserved.

6.3 Who Came First? The Common or the Rare?

It is possible that the origin of rarity began shortly after the origin of life, the extraordinary event that took place on ancient Earth, from which emerged the last universal common ancestor (LUCA). Due to the abrupt changes that occurred during the Archean (Pester et al. 2010), an early diversification by adaptive events likely transpired on early Earth. Considering the impossibility of finding chemical fossils as old as LUCA, no direct evidence has been found to corroborate this diversification event, so, logic has to take the place of evidence to fill the void in data.

Following such logic, the idea that rare species have a key function in regulation, survival, biogeochemical coupling, and community assemblages across all ecosystems may offer indirect evidence for the role that rarity may have had in the early events of life's evolution. These key-stone roles of the rare biosphere support the hypothesis that species rarity has an important role in the temporal stability of the long-term community assemblage (De Anda et al. 2018; Pepe-Ranney et al. 2012).

To elucidate the origin of the rare biosphere in the structuring of primitive communities, analogous communities must be studied. Thus, the study of microbial mats and stromatolites is an excellent approach to recapitulate the assembly of these ancestral niches. These complex communities are perhaps the oldest niches of which there is fossil evidence, showing evidence of life from more than 3.7 billion years ago (Brasier et al. 2002; Knoll et al. 2016).

Modern stromatolites and microbial mats contain finely structured communities of specialized organisms organized by micro-regions and functions. In the outermost layers of these structures live photosynthetic organisms, such as microalgae, cyanobacteria, and diatoms, while in inner layers live anoxygenic photosynthetic bacteria, sulfate-reducing bacteria, and methanogenic archaea. While in the deepest layers, there are a large variety of heterotrophic and anaerobic microorganisms (Wong et al. 2015). Stromatolites and microbial mats coexist with a great amount of diversity represented by the rare biosphere, and this seems to have been vital in giving them stability for millions of years, as they coped with great environmental changes. In this way, they probably worked as diversity hot spots and seed banks during catastrophic events for life (De Anda et al. 2018; Reeder and Knight 2009). Further studies are needed to determine the role of rare species in ancestral models, such as microbialites or microbial mats, to determine if rarity is the product of biotic/abiotic interactions or if rarity is the main determinant that gives stability to communities over long periods of time (Ruiz-González et al. 2019).

6.4   Rare Biosphere and Extremophiles: The Importance of Ancestral Niches for Astrobiological Studies

Today, through technological innovations, life has been found in environments previously considered uninhabitable (Singh et al. 2019). The microorganisms that inhabit these extreme environments are adapted to them, surviving conditions such as extreme temperatures (Galand et al. 2009; Goordial et al. 2016; Russell and Fukunaga 1990), high salt concentrations (Ventosa et al. 2015), low nutrient availability (Plümper et al. 2017; Drummond et al. 2015), high atmospheric pressure, or abnormal pH (Gaboyer et al. 2019; Florentino et al. 2016), among others. For astrobiology, these adaptations are particularly interesting, as they extend the possibilities of detecting life elsewhere in the cosmos.

The first evidence of life on Earth was found in the traces of isotopically light carbon in 4.1 billion-year-old rocks (Tashiro et al. 2017; Dodd et al. 2017); soon thereafter, fossils of microbial communities in stromatolites were discovered. In the case of astrobiology, the ancestral niches built by microbial communities that form stromatolites and microbial mats are a study model of interest (Merino et al. 2019; Clements 1936), especially because these communities are currently found in extreme conditions where life can develop slowly, without the fast-growing algae shading the sun from the microbial phototrophs that are part of such communities. It is, therefore, possible to imagine that ancestral species were likely more similar to the communities of current extremophilic microorganisms, as similar environmental conditions existed on ancient Earth (Rampelotto 2013).

Since the only experiment of life that we know is on Earth, we assume that at the beginning of life, the organisms were associated to environmental conditions as high concentrations of salts, highly variable temperatures, high incidence of UV rays, and a lack of easily accessible nutritional/energy sources (Anbar and Knoll 2002). These conditions led organisms to specialize toward different carbon sources to survive, which subsequently caused a peak in diversification in the highly oligotrophic conditions of early life (Souza et al. 2012). For this reason, it is important to understand the modification and superposition of niches associated with historical, biotic, and abiotic factors, as well as the interactions that determine the abundance of a species in a given habitat.

With this in mind, there is a principle called resource allocation that is based on the costs of staying alive; any organism is limited by its resources, so its survival depends on how it uses resources for its maintenance cycle, growth, and reproduction (Sánchez-Silva et al. 2005). This principle is closely related to the adaptations acquired by extremophiles. Among these examples are thermophiles with specialized membrane lipids that make them more compact and resistant to denaturation at high temperatures (Stetter 1999), as well as psychrophiles who avoid cell freezing by utilizing a greater amount of unbranched fatty acids, an antifreeze intracellular fluid called DMC, and adaptations in metabolic regulation (D'Amico et al. 2006). Acidophiles, on the other hand, employ sulfate as an electron acceptor, as is the case in acid mine drainage and hydrothermal vents, maintaining homeostasis by ion pumping to regulate internal cell pressure with respect to their external environment (Mosier et al. 2013). Similarly, halophiles are osmotically protected with organic solutes, constantly keeping their cell membrane hydrated (Oren 2002). Extremophile studies have been carried out in various extreme environments on Earth, such as in Alaska, Antarctica, hydrothermal vents, hot springs, the ocean floor and its associated faults, and hypersaline lakes in Australia, among other model ecosystems (Galand et al. 2009; Goordial et al. 2016; Plümper et al. 2017; Salas et al. 2015). However, all that is known about the physiology of these extremophilic organisms is from a few cultivated strains; the role of the diverse rare biosphere in these extreme ecosystems largely remains unknown.

The interest of astrobiology in the rare biosphere is centered on the fact that many of the celestial bodies of which their environmental conditions have been described (e.g., the moon, Mars, Titan, and numerous asteroids) have extreme environments potentially habitable by organisms similar to those found among the rare terrestrial biosphere. Mars is particularly interesting as there is evidence that a few billion years ago its environmental conditions were very similar to those of early Earth (Zhang et al. 1993). So far, although the missions to Mars have focused more on chemical description rather than the search for life, the abiotic components discovered, such as the presence of surface solid water, underground liquid, and methane (a simple molecule that on Earth is the product of methanogenesis of Archaea), support the idea of an ancestral or recent existence of extraterrestrial microbial life (Vago et al. 2017). Another candidate for this type of life is Titan, Saturn's largest moon, which has a dense and nitrogen-rich atmosphere. Titan's atmosphere is similar to Earth's in terms of its prevalence of nitrogen, and beneath the atmosphere, the moon likely has methane, ethane, and other hydrocarbons in the liquid state – icy and frozen on the surface but potentially warm in the layers closest to internal magmatic systems. This points to the presence of both atmospheric and surface components that could provide sufficient raw material for biosynthesis, enabling the possibility of an independent and persistent origin of life on Titan (Clarke and Ferris, 1997). On the contrary, even on planets without an atmosphere or with chemically unfavorable atmospheres for life, it is hypothesized that we could still discover microorganisms inhabiting their inner layers.

For the aforementioned reasons, among the most important discoveries for astrobiology in recent times are those that describe microbial life forms hidden in the depths of the Earth that thrive in conditions independent of those in the surface biosphere. According to Belilla and coauthors (2019), new organisms are found every year in environments that were previously thought to be uninhabitable: such as water systems within gold mines with estimated ages of tens of millions of years (Lin et al. 2006); hydrothermal vents deep in the sea (Sogin et al. 2006); lagoons with high temperatures, acidity, or alkalinity (Ward et al. 2006); hypersaline and oligotrophic lagoons (Souza et al. 2012); and even in places in particular the meteoric alkaline water that emanates from a well intersecting quartzite fractures kilometers underground (Moser et al. 2005). In such unique systems, it seems that finally the limits of life have been reached (Belilla et al. 2019) – environments where only organisms with unique abilities can inhabit, isolated and independent of Earth's biosphere and sunlight suggest the possibility of finding homologous isolated biospheres on other planets, proliferating despite hostile surface conditions.

However, access to samples and avoiding contamination represent some of the bigger challenges for the study of these microorganisms (Lin et al. 2006). In addition, being so adapted or specialized to a specific environment with unique metabolisms and interactions makes traditional isolation and culture methods impossible (Al-Awadhi et al. 2013; Zapka et al. 2017). It was not until the development of sequencing technologies and the emergence of metagenomic analysis methodologies that it was possible to study the rare biosphere in greater depth (Lynch and Neufeld 2015).

6.5   Cuatro Ciénegas, a Model for the Study of Rare Biosphere and Astrobiology

The extraordinary oasis in the Cuatro Ciénegas Basin (CCB) is particularly diverse (De Anda et al. 2018; Taboada et al. 2018), not only because each site that has been explored is unique with a large beta diversity (Souza et al. 2006), but also because most of its microbial diversity is part of the rare biosphere, giving each microbial mat, stromatolite, sediment sample, water sample, or soil sample a unique signature (Lee et al. 2017; Souza et al. 2018). The reason behind such a high diversity is its ancestry and unique geological history, making CCB a true lost world that survived different mass extinction events throughout Earth's history. Yet, today in the Anthropocene, CCB risks destruction, as the wetlands are being drained for local agricultural use (Souza et al. 2018).

This "lost world" characteristic is not merely a metaphor; it is of great astrobiological relevance. The ancestral lineages survived in these extremely oligotrophic conditions, which allowed ancient microbes to co-evolve along with their community, resulting in an extraordinary codependency. For this reason, they survived: they could grow and reproduce given the "extended genotype" of the community. It is, therefore, possible that this feature can explain the resilience of life on Earth, helping us explore age-old questions: Does life exist on other planets or the moon? Perhaps, in other planetary systems? If so, how did it survive and diversify? It is highly probable that the survival of the fittest is a universal rule; however, the fittest, in this case, is the one that can coexist and collaborate, not the one that outcompetes the neighbor. May this be a lesson of survival for us all.

6.6 Conclusions

New sequencing technologies, such as the sequencing of highly variable regions of the 16S rRNA gene and shotgun metagenomics, have been crucial in understanding the ecology of microbial communities today. The use of these tools has made it increasingly possible to demonstrate the presence of rare species in communities, identify them, and understand their role within the community. These molecular tools are the reasons why the growth and contributions of research in this topic have led to a rethinking of the first hypotheses about microbial communities and their rare biosphere.

This rare fraction includes a large number of taxa, often exceeding those that dominate the community, maintaining a consistent composition despite its low abundance. There are, however, taxa that are unique to a site, providing a new collection of lineages. This rarity exists mainly due to the particular qualities of the habitat to which it belongs, as is the case with many extremophiles, or even the exclusivity of the biogeographic region where they are found, as is the case of several microorganisms in Cuatro Ciénegas. Regardless of the type of rarity being studied, there is still much to learn about the role of these organisms in microbial communities, as well as the evolutionary and ecological mechanisms that lead a species to be considered rare.

From a functional point of view, the role and importance of the rare biosphere in ecological niches warrant investigation. From the advantages for a species to remain at a low frequency to speculating on the reason why a community maintains its rare biosphere, there are many interesting avenues to explore further.

Thanks to the large amount of diversity found in a set of rare species, one of their most important roles is to function as a seed bank. Together, this pool of species conserves functions that may be necessary during an environmental change, ensuring the survival of the community. This, in turn, can increase the frequency of non-abundant species, making them part of the common biosphere, which creates a dynamic exchange between the abundant and rare. On the other hand, there are some occasions where the rare biosphere holds particular functions required in low abundance, which may play a key role in the regulation, assembly, and survival of the community. This dynamic exchange between rare and abundant, along with the presence of indispensable functions for community assembly and conservation, add a further degree of complexity to the interactions of the community, giving rise to new questions about the structure of microbial communities.

The diversity and ability to survive in extreme conditions make the rare biosphere a good study model for astrobiology. Through its role as a seed bank of metabolic functions, the rare biosphere offers functions of vital importance, such as the degradation of hydrocarbons or the fixation of other carbon sources. In this way, metabolic diversity enables their survival in extreme conditions, such as at high or low temperatures, high levels of salinity, or nutrient shortages. Additionally, these types of conditions are those that most closely resemble a primitive environment on Earth or the environments of other planets. Because of this, the rare biosphere is an excellent study model for investigating the origins of life on Earth, as well as the possibility of life emerging elsewhere.

One hypothesis about the first forms of life after LUCA, is that life could continue due to the rapid diversification that filled the many available niches. It is possible that at the beginning of this diversification, there were no abundant species; rather, there was a large diversity of functions and a low population abundance. In this way, it is parsimonious to suggest that at the beginning, there were generalist organisms that performed different functions, that rapidly evolve into the current diversity we see today. However, given that there is a diversity of functions and most species are specialists, the possibility of their survival alone is lower, even when considering that there were likely low levels of competition. The rare biosphere can then be closely linked to life in a community, with high levels of cooperation. In this way, specialists were able to cover all types of nutrient cycling and be able to survive together as a community, despite abrupt environmental changes suffered by Earth throughout its various geological ages. Thus, evolving communities in stromatolites or microbial mats may have been the key to the survival of life. Similarly, microbes that did not live in communities could have survived because they took refuge in places where their ancestral niches were conserved, as can be observed in hydrothermal vents, deep rocks, blocks of permafrost, and more. These microorganisms are extremophiles, which due to the specificity of the habitat in which they are found are considered rare.

Acknowledgments This project was funded by PAPITT-DGPA grant IG200319 to VS and LEW. We want to thank Laura Espinosa-Asuar and Erika Aguirre-Planter for technical assistance.

References

Aanderud ZT, Jones SE, Fierer N, Lennon JT (2015) Resuscitation of the rare biosphere contributes to pulses of ecosystem activity. Front Microbiol 6:24

Al-Awadhi H, Dashti N, Khanafer M, Al-Mailem D, Ali N, Radwan S (2013) Bias problems in culture-independent analysis of environmental bacterial communities: a representative study on hydrocarbonoclastic bacteria. Springerplus 2(1):369

Anbar AD, Knoll AH (2002) Proterozoic Ocean chemistry and evolution: a bioinorganic bridge? Science 297:1137–1142

Belilla J, Moreira D, Jardillier L, Reboul G, Benzerara K, López-García JM, Bertolino P, López-Archilla AI, López-García P (2019) Hyperdiverse archaea near life limits at the polyextreme geothermal Dallol area. Nat Ecol Evol 3:1552–1561

Bowen JL, Crump BC, Deegan LA, Hobbie JE (2009) Salt marsh sediment bacteria: their distribution and response to external nutrient inputs. ISME J 3:924–934

Brasier MD, Green OR, Jephcoat AP, Kleppe AK, Van Kranendonk MJ, Lindsay JF, Steele A, Grassineau NV (2002) Questioning the evidence for Earth's oldest fossils. Nature 416(6876):76–81

Clarke DW, Ferris JP (1997) Chemical evolution on Titan: comparisons to the prebiotic earth. Orig Life Evol Biosph 27:225–248

Clements FE (1936) Nature and structure of the climax. J Ecol 24:252

Coveley S, Elshahed MS, Youssef NH (2015) Response of the rare biosphere to environmental stressors in a highly diverse ecosystem (Zodletone spring, OK, USA). Peer J 2015

D'Amico S, Collins T, Marx JC, Feller G, Gerday C (2006) Psychrophilic microorganisms: challenges for life. EMBO Rep 7(4):385–389

De Anda V, Zapata-Peñasco I, Blaz J, Poot-Hernandez AC, Contreras-Moreira B, Hernandez Rosales M, Eguiarte LE, Souza V (2018) Understanding the mechanisms behind the response to environmental perturbation in microbial mats: a metagenomic-network based approach. Front Microbiol 9:2606

Dodd MS, Papineau D, Grenne T, Slack JF, Rittner M, Pirajno F, O'Neil J, Little CT (2017) Evidence for early life in Earth's oldest hydrothermal vent precipitates. Nature 543:60–64

Drummond JBR, Pufahl PK, Porto CG, Carvalho M (2015) Neoproterozoic peritidal phosphorite from the Sete Lagoas formation (Brazil) and the Precambrian phosphorus cycle. Sedimentology 62:1978–2008

Dybdahl MF, Lively CM (1998) Host-parasite coevolution: evidence for rare advantage and time-lagged selection in a natural population. Evolution 52(4):1057–1066

Elshahed MS, Youssef NH, Spain AM, Sheik C, Najar FZ, Sukharnikov LO, Roe BA, Davis JP, Schloss PD, Bailey VL, Krumholz LR (2008) Novelty and uniqueness patterns of rare members of the soil biosphere. Appl Environ Microbiol 74:5422–5428

Fetzer I, Johst K, Schäwe R, Banitz T, Harms H, Chatzinotas A (2015) The extent of functional redundancy changes as species' roles shift in different environments. Proc Natl Acad Sci U S A 112(48):14888–14893

Florentino AP, Weijma J, Stams AJM, Sánchez-Andrea I (2016) Ecophysiology and application of acidophilic sulfur-reducing microorganisms. In: Rampelotto PH (ed) Biotechnology of extremophiles. Springer, Cham, pp 141–175

Franklin RB, Garland JL, Bolster CH, Mills AL (2001) Impact of dilution on microbial community structure and functional potential: comparison of numerical simulations and batch culture experiments. Appl Environ Microbiol 67:702–712

Fuentes S, Méndez V, Aguila P, Seeger M (2014) Bioremediation of petroleum hydrocarbons: catabolic genes, microbial communities, and applications. Appl Microbiol Biotechnol 98:4781–4794

Gaboyer F, Burgaud G, Edgcomb V (2019) The deep subseafloor and biosignatures. In: Cavalazzi B, Westall F (eds) Biosignatures for Astrobiology. Springer, Cham, pp. 87–109

Galand PE, Casamayor EO, Kirchman DL, Lovejoy C (2009) Ecology of the rare microbial biosphere of the Arctic Ocean. Proc Natl Acad Sci U S A 106:22427–22432

Gobet A, Böer SI, Huse SM, Van Beusekom JE, Quince C, Sogin ML, Boetius A, Ramette A (2012) Diversity and dynamics of rare and of resident bacterial populations in coastal sands. ISME J 6(3):542–553

Goordial J, Davila A, Lacelle D, Pollard W, Marinova MM, Greer CW, DiRuggiero J, McKay CP, Whyte LG (2016) Nearing the cold-arid limits of microbial life in permafrost of an upper dry valley, Antarctica. ISME J 10(7):1613–1624

Griffiths BS, Kuan HL, Ritz K, Glover LA, McCaig AE, Fenwick C (2004) The relationship between microbial community structure and functional stability, tested experimentally in an upland pasture soil. Microb Ecol 47(1):104–113

Hausmann B, Pelikan C, Rattei T et al (2019) Long-term transcriptional activity at zero growth of a cosmopolitan rare biosphere member. MBio 10

Hernandez-Raquet G, Durand E, Braun F, Cravo-Laureau C, Godon JJ (2013) Impact of microbial diversity depletion on xenobiotic degradation by sewage-activated sludge. Environ Microbiol Rep 5:588–594

Kazancıoğlu E, Arnqvist G (2014) The maintenance of mitochondrial genetic variation by negative frequency-dependent selection. Ecol Lett 17:22–27. https://doi.org/10.1111/ele.12195

Knoll AH, Bergmann KD, Strauss JV (2016) Life: the first two billion years. Philos Trans R Soc B Biol Sci 371(1707):20150493

Lee ZM, Poret-Peterson AT, Siefert JL, Kaul D, Moustafa A, Allen AE, Dupont CL, Eguiarte LE, Souza V, Elser JJ (2017) Nutrient stoichiometry shapes microbial community structure in an evaporitic shallow pond. Front Microbiol 8:949

Lin LH, Wang PL, Rumble D, Lippmann-Pipke J, Boice E, Pratt LM, Lollar BS, Brodie EL, Hazen TC, Andersen GL, DeSantis TZ (2006) Long-term sustainability of a high-energy, low-diversity crustal biome. Science 314(5798):479–482

Locey KJ, Lennon JT (2016) Scaling laws predict global microbial diversity. Proc Natl Acad Sci U S A 113:5970–5975

Lynch MDJ, Neufeld JD (2015) Ecology and exploration of the rare biosphere. Nat Rev Microbiol 13:217–229

Mallon CA, Poly F, Le Roux X, Marring I, van Elsas JD, Salles JF (2015) Resource pulses can alleviate the biodiversity-invasion relationship in soil microbial communities. Ecology 96:915–926

Matos A, Kerkhof L, Garland JL (2005) Effects of microbial community diversity on the survival of Pseudomonas aeruginosa in the wheat rhizosphere. Microb Ecol 49:257–264

Merino N, Aronson HS, Bojanova DP, Feyhl-Buska J, Wong ML, Zhang S, Giovannelli D (2019) Living at the extremes: extremophiles and the limits of life in a planetary context. Front Microbiol 10:780

Minter EJA, Watts PC, Lowe CD, Brockhurst MA (2015) Negative frequency-dependent selection is intensified at higher population densities in protist populations. Biol Lett 11:20150192

Moser DP, Gihring TM, Brockman FJ et al (2005) Desulfotomaculum and Methanobacterium spp. dominate a 4- to 5-kilometer-deep fault. Appl Environ Microbiol 71:8773–8783

Mosier AC, Justice NB, Bowen BP et al (2013) Metabolites associated with adaptation of microorganisms to an acidophilic, metal-rich environment identified by stable-isotope-enabled metabolomics. MBio 4:e00484-12

Musat N, Halm H, Winterholler B, Hoppe P, Peduzzi S, Hillion F, Horreard F, Amann R, Jørgensen BB, Kuypers MM (2008) A single-cell view on the ecophysiology of anaerobic phototrophic bacteria. Proc. Natl. Acad. Sci USA 105:17861–17866

Nemergut DR, Costello EK, Hamady M, Lozupone C, Jiang L, Schmidt SK, Fierer N, Townsend AR, Cleveland CC, Stanish L, Knight R (2011) Global patterns in the biogeography of bacterial taxa. Environ Microbiol 13:135–144

Oren A (2002) Diversity of halophilic microorganisms: environments, phylogeny, physiology, and applications. J Ind Microbiol Biotechnol 28(1):56–63

Pedrós-Alió C (2012) The rare bacterial biosphere. Annu Rev Mar Sci 4:449–466

Pepe-Ranney C, Berelson WM, Corsetti FA, Treants M, Spear JR (2012) Cyanobacterial construction of hot spring siliceous stromatolites in Yellowstone National Park. Environ Microbiol 14(5):1182–1197

Pester M, Bittner N, Deevong P, Wagner M, Loy A (2010) A 'rare biosphere' microorganism contributes to sulfate reduction in a peatland. ISME J 4(12):1591

Philippot L, Raaijmakers JM, Lemanceau P, Van Der Putten WH (2013) Going back to the roots: the microbial ecology of the rhizosphere. Nat Rev Microbiol 11:789–799

Plümper O, King HE, Geisler T, Liu Y, Pabst S, Savov IP, Rost D, Zack T (2017) Subduction zone forearc serpentinites as incubators for deep microbial life. Proc Natl Acad Sci U S A 114:4324–4329

Rampelotto PH (2013) Extremophiles and extreme environments. Life 3:482–485

Reeder J, Knight R (2009) The "rare biosphere": a reality check. Nat Methods 6:636–637

Rousk J, Brookes PC, Bååth E (2009) Contrasting soil pH effects on fungal and bacterial growth suggest functional redundancy in carbon mineralization. Appl Environ Microbiol 75:1589–1596

Ruiz-González C, Logares R, Sebastián M, Mestre M, Rodríguez-Martínez R, Galí M, Sala MM, Acinas SG, Duarte CM, Gasol JM (2019) Higher contribution of globally rare bacterial taxa reflects environmental transitions across the surface ocean. Mol Ecol 28:1930–1945

Russell NJ, Fukunaga N (1990) A comparison of thermal adaptation of membrane lipids in psychrophilic and thermophilic bacteria. FEMS Microbiol Lett 75:171–182

Salas EC, Bhartia R, Anderson L et al (2015) In situ detection of microbial life in the deep biosphere in igneous ocean crust. Front Microbiol 6:1260

Sánchez-Silva M, Daniels M, Lleras G, Patino D (2005) A transport network reliability model for the efficient assignment of resources. Transp Res Part B Methodol 39:47–63

Scholz MB, Lo C-C, Chain PSG (2012) Next generation sequencing and bioinformatic bottlenecks: the current state of metagenomic data analysis. Curr Opin Biotechnol 23:9–15

Shade A, Peter H, Allison SD, Baho D, Berga M, Bürgmann H, Huber DH, Langenheder S, Lennon JT, Martiny JB, Matulich KL (2012) Fundamentals of microbial community resistance and resilience. Front Microbiol 3:417

Singh P, Jain K, Desai C, Tiwari O, Madamwar D (2019) Microbial community dynamics of extremophiles/extreme environment. In: Das S, Dash HR, Microbial diversity in the genomic era. Academic Press, London, pp 323–332

Sogin ML, Morrison HG, Huber JA, Welch DM, Huse SM, Neal PR, Arrieta JM, Herndl GJ (2006) Microbial diversity in the deep sea and the underexplored "rare biosphere.". Proc Natl Acad Sci 103:12115–12120

Souza V, Espinosa-Asuar L, Escalante AE, Eguiarte LE, Farmer J, Forney L, Lloret L, Rodríguez-Martínez JM, Soberón X, Dirzo R, Elser JJ (2006) An endangered oasis of aquatic microbial biodiversity in the Chihuahuan desert. Proc Natl Acad Sci U S A 103:6565–6570

Souza V, Siefert JL, Escalante AE, Elser JJ, Eguiarte LE (2012) The Cuatro Ciénegas basin in Coahuila, Mexico: an astrobiological Precambrian park. Astrobiology 12(7):641–647

Souza V, Moreno-Letelier A, Travisano M, Alcaraz LD, Olmedo G, Eguiarte LE (2018) The lost world of Cuatro Cienegas Basin, a relictual bacterial niche in a desert oasis. elife 7:e38278

Stetter KO (1999) Extremophiles and their adaptation to hot environments. FEBS Lett 452:22–25

Taboada B, Isa P, Gutiérrez-Escolano AL, Del Ángel RM, Ludert JE, Vázquez N, Tapia-Palacios MA, Chávez P, Garrido E, Espinosa AC, Eguiarte LE (2018) The geographic structure of viruses in the Cuatro Ciénegas Basin, a unique oasis in northern Mexico, reveals a highly diverse population on a small geographic scale. Appl Environ Microbiol 84(11):e00645–e00618

Tashiro T, Ishida A, Hori M, Igisu M, Koike M, Méjean P, Takahata N, Sano Y, Komiya T (2017) Early trace of life from 3.95 Ga sedimentary rocks in Labrador, Canada. Nature 549(7673):516–518

Taylor MW, Tsai P, Simister RL, Deines P, Botte E, Ericson G, Schmitt S, Webster NS (2012) 'Sponge-specific' bacteria are widespread (but rare) in diverse marine environments. ISME J 7(2):438

Tilman D (1999) The ecological consequences of changes in biodiversity: a search for general principles. Ecology 80:1455–1474

Vago JL, Westall F, Coates AJ, Jaumann R, Korablev O, Ciarletti V, Mitrofanov I, Josset JL, De Sanctis MC, Bibring JP, Rull F (2017) Habitability on early Mars and the search for biosignatures with the ExoMars Rover. Astrobiology 17(6–7):471–510

van Elsas JD, Chiurazzi M, Mallon CA, Elhottová D, Krištůfek V, Salles JF (2012) Microbial diversity determines the invasion of soil by a bacterial pathogen. Proc Natl Acad Sci U S A 109:1159–1164

Ventosa A, de la Haba RR, Sánchez-Porro C, Papke RT (2015) Microbial diversity of hypersaline environments: a metagenomic approach. Curr Opin Microbiol 25:80–87

Vivant AL, Garmyn D, Maron PA, Nowak V, Piveteau P (2013) Microbial diversity and structure are drivers of the biological barrier effect against *Listeria monocytogenes* in soil. PLoS One 8(10):e76991

Walke JB, Becker MH, Loftus SC, House LL, Cormier G, Jensen RV, Belden LK (2014) Amphibian skin may select for rare environmental microbes. ISME J 8(11):2207–2217

Wang Y, Hatt JK, Tsementzi D, Rodriguez-R LM, Ruiz-Pérez CA, Weigand MR, Kizer H, Maresca G, Krishnan R, Poretsky R, Spain JC (2017) Quantifying the importance of the rare biosphere for microbial community response to organic pollutants in a freshwater ecosystem. Appl Environ Microbiol 83(8):e03321-16

Ward DM, Bateson MM, Ferris MJ, Kühl M, Wieland A, Koeppel A, Cohan FM (2006) Cyanobacterial ecotypes in the microbial mat community of mushroom spring (Yellowstone National Park, Wyoming) as species-like units linking microbial community composition, structure and function. Philos T R Soc B 361(1475):1997–2008

Weeks AR, Hoffmann AA (2008) Frequency-dependent selection maintains clonal diversity in an asexual organism. Proc Natl Acad Sci U S A 105:17872–17877

Winter C, Bouvier T, Weinbauer MG, Thingstad TF (2010) Trade-offs between competition and defense specialists among unicellular planktonic organisms: the "killing the winner" hypothesis revisited. Microbiol Mol Biol Rev 74:42–57

Wong HL, Smith D-L, Visscher PT, Burns BP (2015) Niche differentiation of bacterial communities at a millimeter scale in Shark Bay microbial mats. Sci Rep 5:15607

Zapka C, Leff J, Henley J, Tittl J, De Nardo E, Butler M, Griggs R, Fierer N, Edmonds-Wilson S (2017) Comparison of standard culture-based method to culture-independent method for evaluation of hygiene effects on the hand microbiome. MBio 8(2):e00093–e00017

Zhang MHG, Luhmann JG, Bougher SW, Nagy AF (1993) The ancient oxygen exosphere of Mars: implications for atmosphere evolution. J Geophys Res Planets 98:10915–10923

Zhang Y, Wu G, Jiang H, Yang J, She W, Khan I (2018) Abundant and rare microbial biospheres respond differently to environmental and spatial factors in Tibetan Hot Springs. Front Microbiol 9:2096

Chapter 7
Bacterial Communities from Deep Hydrothermal Systems: The Southern Gulf of California as an Example of Primeval Environments

Laura Espinosa-Asuar, Luis A. Soto, Diana L. Salcedo,
Abril Hernández-Monroy, Luis E. Eguiarte, Valeria Souza, and Patricia Velez

Abstract Deep hydrothermal systems result from the magmatic and tectonic activity of the ocean floor. This deep extreme biosphere represents a unique oasis of life driven by sulfur-based chemosynthesis instead of photosynthesis. The organisms inhabiting these systems are adapted to cope with harsh environmental conditions such as the absence of sunlight, high temperatures and hydrostatic pressures, and elevated concentrations of hydrogen sulfide, as well as high concentrations of heavy metals. Therefore, this biome is different from any other environment on modern Earth. As expected from such conditions, chemoautotrophic prokaryotes are the leading primary producers at the base of the food web considered as an analog to the oldest signs of life on Earth. Herein, we discuss prokaryotic diversity and community structure from the newly discovered hydrothermal systems in the Alarcón Rise (AR), the Pescadero Basin (PB), and the Pescadero Transform Fault (PTF) at the mouth of the Gulf of California, Mexico, using 16S rRNA gene amplicon Illumina sequencing. Despite the spatial proximity of the studied vent systems (<100 km),

L. Espinosa-Asuar (✉) · L. E. Eguiarte · V. Souza
Instituto de Ecología, Universidad Nacional Autónoma de México, UNAM,
Mexico City, Mexico
e-mail: lauasuar@ecologia.unam.mx

L. A. Soto
Instituto de Ciencias del Mar y Limnología, Universidad Nacional Autónoma de México,
UNAM, Mexico City, Mexico

D. L. Salcedo
Posgrado en Ciencias del Mar y Limnología, Universidad Nacional Autónoma de México,
UNAM, Mexico City, Mexico

A. Hernández-Monroy · P. Velez (✉)
Departamento de Botánica, Instituto de Biología, Universidad Nacional Autónoma de
México, UNAM, Mexico City, Mexico
e-mail: pvelez@ib.unam.mx

© Springer Nature Switzerland AG 2020   149
V. Souza et al. (eds.), *Astrobiology and Cuatro Ciénegas Basin as an Analog of
Early Earth*, Cuatro Ciénegas Basin: An Endangered Hyperdiverse Oasis,
https://doi.org/10.1007/978-3-030-46087-7_7

they differ considerably in their physical, chemical, geological settings, and biotic characteristics. Our results indicated that beta prokaryotic diversity is associated to the sampling source, suggesting a strong effect of environmental conditions in shaping microbial distribution. The most abundant phyla were Proteobacteria, Bacteroidetes, Actinobacteria, Chloroflexi, and Epsilonbacteraeota. Also, we found evidence on the oxidation of methane as a prevalent process in PB and PTF, since methylotrophic bacteria and Atribacteria were abundant, in contrast to AR basalt-hosted system. Bacteria associated with the sulfur cycle, in particular sulfur compounds reducing and sulfur compounds oxidizing bacteria predominated in all samples, confirming the importance of sulfur supporting vent communities. It is possible that vent systems played a significant role in the origins of life on Earth. Hence, they represent useful models when searching for life elsewhere in the universe.

7.1 Introduction

Since the hydrothermal systems were discovered in the Galapagos Ridge, ca. 40 years ago (Corliss et al. 1979), they have proved to be a key feature on the ocean seafloor where magmatic and tectonic activity occurs. Presently, more than 720 vent systems are known from mid-ocean ridges and volcanic arcs (Beaulieu and Szafranski 2018). The seawater percolating through porous sediments or basaltic rocks entering in contact with the magmatic chamber originates the emission to the seafloor of hot fluids (>200 °C), containing an enriched mixture of chemical reduced compounds such as sulfide, hydrogen, methane, manganese, and iron (Brazelton 2017; Dick 2019). When these hot fluids come in contact with the cold and well-oxygenated deep water they precipitate, forming complex edifices shaped as chimneys or small spires on the seafloor surface. These hydrothermal fluids represent the chemical energy that sustains one of the most enigmatic forms of marine life found along the plate tectonic boundaries of the ocean floor: the hydrothermal vent communities (Dick 2019). The existence of these life forms that are derived from geothermal energy is considered a good model to search for living conditions in other planets or moons (Barge and White 2017).

Vent communities are epitomized by a unique assemblage of invertebrate species that flourish under extreme environmental conditions of temperature, darkness, pressure, and toxicity (Hessler and Kaharl 1995). These ephemeral biotic assemblages depend strictly on the dynamics of the hydrothermal fluid itself, lacking temporal stability: if the fluid emission ceases or is obliterated by tectonic processes, the supply of chemical energy contained in electron donors such as S^{-2}, Fe^{2+}, H_2, and Mn^{2+} is no longer available (Baker et al. 1989). These invertebrate deep-sea communities are possible due to the unseen majority, i.e., the prokaryotic chemolithoautotrophic microorganisms (Pjevac et al. 2018); these microbial communities constitute the primary producers in the vent community, comprising mesophiles, thermophiles, and hyperthermophiles, generally from Phylum Aquificae, and Class Deltaproteobacteria, Epsilonproteobacteria, and Gammaproteobacteria (Dick 2019).

Hydrothermal vent systems harbor an ecosystem based on microbial communities with reducing power of hydrogen sulfide or dissolved hydrogen gas as a source of electrons (Deamer and Damer 2017). Here, processes associated with serpentinization (where mineral components in the olivine undergo an exothermic reaction that produces hydrogen gas and heat; Kelley et al. 2005) offer further parallels with the biochemistry of ancient autotrophic cells (Herschy et al. 2014): hydrogen and sulfur provide a source of chemical energy from solutes and minerals at different redox states and pH ranges (Wächtershäuser and User 1992; Zierenberg et al. 2000) that probably reflect the most ancient forms of autotrophy. The associated microbial consortium in these extreme environments includes methanogenic and sulfooxidizing prokaryotes (Sievert et al. 2007) that transform inorganic energy resources into organic compounds employing a diversity of complex metabolic pathways to fix CO_2, via at least six known different metabolic pathways (Hügler and Sievert 2011). In nearly all the hydrothermal systems, denitrification processes coupled to sulfur compounds oxidation and hydrogen oxidation play a key role (Bourbonnais et al. 2012, 2014; Li et al. 2018).

As described by Martin et al. (2008), the chemistry of life is the chemistry of reduced organic compounds, and this is particularly true for hydrothermal systems. These chemically reactive environments provide suitable conditions for sustained prebiotic syntheses, and in consequence, researchers interested in the origin of life have paid attention to hydrothermal systems (Corliss et al. 1981; Baross and Hoffman 1985; Russell et al. 1993; Russell and Hall 1997; Zierenberg et al. 2000; Kelley et al. 2001). In particular, the sulfide chimneys of the vent systems are primordial environments evoking early Earth conditions, with reactive gases, dissolved elements, and sharp thermal and chemical gradients (Martin et al. 2008).

In addition, it has been suggested that a hydrothermal vent origin could allow life to begin in the Enceladus Ocean (in the sixth largest moon of Saturn with possible hydrothermal activity), on Europa (moon orbiting Jupiter with possible hydrothermal activity), and even indirectly on early Mars (Hsu et al. 2015). This conjecture has been recently explored (Deamer et al. 2019), leading to a fascinating framework aiming to understand aqueous environments likely to be present on the prebiotic Earth.

In 2000, a milestone was reached with the discovery of the unique hydrothermal field at 700–800 m below the sea surface, the Lost City hydrothermal field (LCHF), which is characterized by carbonate chimneys (Von Damm 2001). This system has been considered as crucial to understanding the origin of life since the ultramafic underpinnings of LCHF have a similar chemical composition to lavas that erupted into the primordial oceans on early Earth (de Wit 1997; Kelley et al. 2001, 2005).

Remarkably, in recent years, a new carbonate vent system was discovered at 3800 m depth in the Pescadero Basin (PB), Southern Gulf of California, Mexico, in the northernmost segments of the East Pacific Rise (EPR) (Paduan et al. 2018). This site harbors an unexplored microbial diversity (with the EPR and Guaymas Basin as the nearest known hydrothermal vents prokaryote communities; Teske et al. 2002; Dhillon et al. 2003; Campbell et al. 2013; He and Zhang 2016; Dombrowski et al. 2018). So, in the present study, we examined the prokaryotic diversity and community structure from hydrothermal systems in the Pescadero Basin (PB) and the neighboring Alarcón Rise (AR), and Pescadero Transform Fault (PTF) systems using a 16S rRNA gene amplicon Illumina sequencing approach to contribute to our knowledge on the diversity and functioning of these prebiotic analog sites.

7.2 Materials and Methods

7.2.1 Study Area

The Gulf of California (also known as Sea of Cortes) is a semi-enclosed basin approximately 1000 km long and 150 km wide on average. It lies between the peninsula of Baja California and the northwestern continental region of Mexico, and it is connected to the Tropical East Pacific at its southern end (Álvarez-Borrego 1983).

There is a series of basins along the gulf, in which the depth increases to the south. The Pescadero Basin (PB) system is located at the entrance of the Gulf, the high-temperature vent field located there is at 3700 m depth, and it is a sedimentary basin with mounds and chimneys composed of white and brown hydrothermal calcite, with heights ranging from 12 to 25 m, and low mounds spread over an area of approximately 0.2–0.5 km (Paduan et al. 2018). End-member fluids contained high concentrations of aromatic hydrocarbons, hydrogen, methane, and hydrogen sulfide. The temperatures reach up to 290 °C, and their pH has been measured at 6.5 (Goffredi et al. 2017; Paduan et al. 2018). The biotic community is dominated by the siboglinid tubeworm *Oasisia aff. alvinae* (Goffredi et al. 2017).

At the southeastern of PB, seepage through sediments and outcrops occurs along the Pescadero Transform Fault (PTF), extending for 60 km. The low-temperature vent systems are located at 23.64° and 2400 m in depth (Goffredi et al. 2017; Salcedo et al. 2019). Only discrete venting was observed in volcanic rubble and sediments, emanating at temperatures of 5°C (Clague et al. 2018), which supports scattered dense patches of the coexisting siboglinid tubeworms of the species *Escarpia spicata* and *Lamellibrachia barhami* (Goffredi et al. 2017). This community is less complex and diverse than the one in PB, as only 15 taxa have been identified and the trophic web seems to be quite short (Salcedo et al. 2019).

The Alarcón Rise (AR) is the northernmost bare-rock segment of the EPR, and it is adjacent to the southeastern tip of Baja California (Clague et al. 2018). This locality hosts young basalt high-temperature vents with black smokers with a maximum temperature of 360 °C. The fluids have high concentrations of hydrogen sulfide, little to no hydrogen, methane, or hydrocarbons. The chimneys formed by hydrothermal fluids are polymetallic sulfide deposits. The dominant species is the siboglinid tubeworm *Riftia pachyptila* (Goffredi et al. 2017).

7.2.2   Sampling

The sediment samples were taken during the expedition Vents and Seeps conducted in 2015 by Monterey Bay Aquarium Research Institute (MBARI), in the Southern Gulf of California onboard the R/V Western Flyer, equipped with the remotely operated vehicle (ROV) Doc Ricketts. The sampling was performed using short push cores (30 cm) operated by the articulating manipulator arm of the ROV at easy access randomly chosen sites. Samples were taken from three different deep-sea extreme ecosystems: high-temperature vents in PB (three cores) and AR (one core), and low-temperature vents at PTF (two cores). Surficial sediments subsamples (10 cm depth) were obtained from the push cores and deep- frozen at −80 °C.

7.2.3   Total DNA Isolation, Construction of 16S rRNA Gene Libraries, and Illumina Sequencing

The entire community DNA was isolated from six sediment core samples collected in the three studied vent systems (Fig. 7.1), using the FastDNA®SPIN kit (MP Biomedicals, Irvine, CA) for soils, according to the manufacturer's instructions. An approximately 280-bp of the V3-V4 region of the 16S rRNA gene from 357F primer

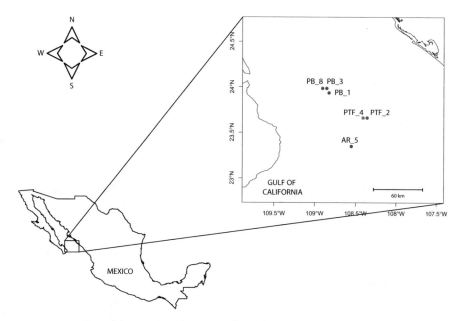

Fig. 7.1 Location of the three studied deep hydrothermal vent systems in the southern Gulf of California, Mexico

(Rudi et al. 1997) was commercially amplified and sequenced using the Illumina MiSeq 300 platform at the Laboratorio de Servicios Genómicos, LANGEBIO.

7.2.4   Bioinformatic and Phylogenetic Analyses of 16S rRNA Amplicon Data

Single End (SE) forward sequences were quality filtered and dereplicated using the QIIME 2 bioinformatics platform version 2019.7 (Bolyen et al. 2018). Initial sequences were demultiplexed and truncated at the 10th bp from the left, and the 280th bp from the right. Divisive Amplicon Denoising Algorithm 2 (DADA2) was then used to filter by sequence quality, denoise, create a sample/operational taxonomic units (OTU) table, and remove chimeras. OTUs were then taxonomically assigned using the Silva-132-99-nb-classifier database. Those OTUs that occurred <10 times across all six samples were removed. The final OTU table was rarefied to 20,000 sequences to make unbiased diversity comparisons and used for all subsequent analyses.

QIIME 2 was then used to obtain alpha and beta biodiversity scores. To test the correlation between the Jaccard Dissimilarity Index values and geographical distances, a Mantel test was performed using Mantel Function as implemented in Vegan R library v1.42 R (Oksanen et al. 2013), and the UpSet plot was prepared by R library UpSetR (Conway et al. 2017).

7.3   Results and Discussion

7.3.1   Alpha Prokaryotic Diversity Patterns

The analysis generated 564,117 reads after quality trimming and a chimera check, comprising 7818 OTUs, with a range of 44,368–121,337 reads per site. QIIME2 taxonomic assignment showed 60 different bacterial phyla (Fig. 7.2), a high number in relation to the adjacent hydrothermal vent system in Guaymas Basin where 29 bacterial phyla were reported (Dombrowski et al. 2018). These diversity results also surpass those reported for the East Pacific Rise with 37 bacterial phyla (He and Zhang 2016). Overall, this suggests that vent fields in the southern Gulf of California are diverse systems in terms of microbial communities, but this statement should be taken with caution since these studies used different methodological approaches (e.g., genomes and 454 pyrosequencing respectively).

The most abundant phylum among all the samples was Proteobacteria (Fig. 7.2), representing 45% of all samples. In particular, Delta- and Gamma-, and Epsilonproteobacteria – recently reclassified to a new phylum, Epsilonbacteraeota (Waite et al. 2017) – were abundant in our samples. This abundance pattern

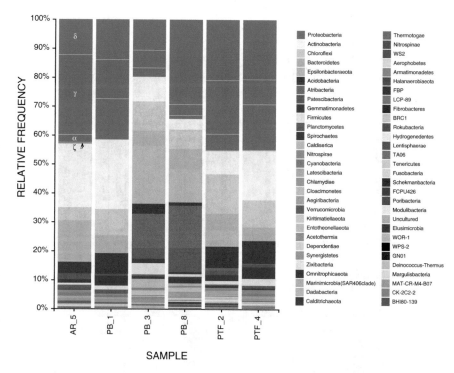

Fig. 7.2 Bacterial communities inhabiting the studied deep hydrothermal systems in the Gulf of California. Histogram depicts phyla richness and relative frequency within samples collected from Alarcón Rise (AR), Pescadero Basin (PB), and Pescadero Transform Fault (PTF) (α = Alphaproteobacteria, γ = Gammaproteobacteria, δ = Deltaproteobacteria, ζ = Zetaproteobacteria)

coincided with previous studies reporting these taxa as a major component of pro-karyotic communities in hydrothermal vents (Dick 2019), particularly, in similar systems, such as the Guaymas Basin (Teske et al. 2002; Dombrowski et al. 2018).

Actinobacteria, Chloroflexi, and Bacteroidetes represented the second most abundant phyla in our samples, again similar to bacterial community composition in vent systems in Guaymas Basin and East Pacific Rise (Teske et al. 2002; Campbell et al. 2013; He and Zhang 2016; Dombrowski et al. 2018). In particular, Actinomarinales OTUs were observed to be a dominant component in all samples (Table 7.1). This group of bacteria has been regarded as common marine bacteria from the photic zone, being the smallest free-living cells among prokaryotes (Ghai et al. 2013). Nonetheless, to the extent of our knowledge, this group has not been reported from hydrothermal vents previously, reflecting the ample capacity of marine microorganisms to adapt and colonize extreme environments.

The third most abundant phyla were Acidobacteria, Artribacteria, and Patesibacteria that have also been detected in the neighboring vent system of Guaymas Basin (Teske et al. 2002; Dombrowski et al. 2018), despite being considered as a rare component of active hydrothermal vents (Zhang et al. 2016).

Table 7.1 List of shared OTUs within samples from hydrothermal vent systems in the southern Gulf of California

TAXA	PTF_2	PTF_4	PB_1	PB_3	PB_8	AR_5	POTENTIAL ROLE / CHARACTERISTICS
Actinobacteria; Actinomarinales order (10 OTUs)	1.8	5.3	8	2.5	1.1	14.9	Smalest free living cells (Ghai et al. 2013)
Atribacteria; JS1 class (4 OTUs)	0	0	0	11	14.2	0	Fermentation products that support methanogens (Carr et al. 2015)
Alphaproteobacteria; *Methyloceanibacter* genus (1 OTU)	0.4	1	0.6	0.03	0.05	0.03	Ch4 oxidation (Takeuchi et al. 2014)
Epsilonbacteraeota; *Sulfurovum* genus (2 OTUs)	2.6	0	0.05	18.2	10.3	0.06	Sulfur compounds oxidation / nitrogen cycle (Mori et al. 2018)
Gammaproteobacteria; uncultured *Thioalkalivibrio sp.* (1 OTU)	0.1	0	0	0	0	5.9	Sulfur compounds oxidation (Cao et al. 2014)
Deltaproteobacteria; Desulfobacterales order (1 OTU)	0	0	0	3.6	17	0	Sulfur compounds reduction (Campbell et al. 2013)
Deltaproteobacteria; Desulfobacterales order; Sva0081 sediment group genus (1 OTU)	0.4	0.5	0.2	0.006	0	0	Sulfur compounds reduction (Campbell et al. 2013)
Deltaproteobacteria; Desulfobacterales order; *Desulfobulbus* genus (1 OTU)	0	0	0	0	0	1.8	Sulfur compounds reduction (Campbell et al. 2013)
Deltaproteobacteria; Syntrophobacterales order; Syntrophobacteraceae family (1 OTU)	0.9	2.4	0.4	0.005	0.004	0.007	Sulfur compounds reduction (Kuever et al. 2005)
Synergistetes; *Thermovirga* genus (1 OTU)	0	0	0	1	0	0	Isolated from an oil well (Dahle and Birkeland 2006)
Cloacimonetes; MSBL8 family (1 OTU)	0	0	0	1.1	0.5	0	Long fatty chains degradation (Shakeri Yekta et al. 2019)
Deltaproteobacteria; *Deferrisoma* genus (1 OTU)	0	0	0	0	0	0.1	Iron metabolism (Slobodkina et al. 2012)
Gammaproteobacteria; Acidiferrobacteraceae family (1 OTU)	0	0	0	0	0	0.08	Iron metabolism (Issotta et al. 2018)
Zetaproteobacteria; *Mariprofundus* genus (1 OTU)	0	0	0	0	0	0.3	Iron metabolism (Emerson et al. 2007)

Results reflect samples collected from Alarcón Rise (AR), Pescadero Basin (PB), and Pescadero Transform Fault (PTF). Values (occurring in more than in a single site), abundant (≥1% reads) and iconic taxa (based on their reported metabolic capacities) within the overall OTUs database in percentage are categorized into three levels of color intensity (<0.09%; <0.9%; >1%)

Overall, these diversity patterns denote a marked similarity in terms of the dominant component of high bacterial taxonomic ranks with neighboring vent systems, which may suggest a significant effect of the environmental conditions in the assemblage vent microbial communities.

At a local level, alpha diversity estimates showed that PTF sites and a single sample from PB (PB_1) were the most diverse locations within the studied vent systems (Table 7.2). Also, these sites showed high diversity Shannon values (9.3–10) even when compared to hydrothermal vents in the East Pacific Rise and Guaymas Basin (Campbell et al. 2013; He and Zhang 2016). The lowest diversity values were obtained from PB samples (6.6, 7.6), and the AR showed intermediate values (8.3). Interestingly, the opposite pattern has been reported for macrofauna (e.g., *E. spicata* and *L. barhami*), since PTF harbors the lowest diversity levels in contrast to the PB system (Goffredi et al. 2017). This result may be linked to the ability of microorganisms to occupy micro-niches in relation to the macrofauna, contributing substantially to ecosystem functions (Troussellier et al. 2017).

7.3.2 Beta Prokaryotic Diversity Patterns

Beta diversity distribution showed a strong association between the sample sources (Fig. 7.3). This is related to each hydrothermal system exhibiting dissimilar physicochemical conditions (Goffredi et al. 2017; Paduan et al. 2018; Salcedo et al. 2019). Consequently, samples from the same locality were clustered, except for PB_1 that grouped to the PTF. Therefore, the Mantel test showed no significance for the geographic distribution. These results suggest a strong effect of environmental conditions in shaping microbial distribution, given the significant differences in physicochemical, geological, and biotic settings among the three analyzed localities (PB, PTF, and AR; Goffredi et al. 2017; Paduan et al. 2018; Salcedo et al. 2019).

To attain a better understanding of beta distribution patterns, diversity was explored at the OTUs level (Fig. 7.4, Table 7.2). This approach demonstrated that the majority of OTUs was unique to each sediment core (45.6% to 79.7% of OTUs, Fig. 7.4), embodying a large number of rare species of microbial life known as the

Table 7.2 Observed OTUs and alpha diversity values (Faith phylogenetic diversity (PD) and Shannon), within the six samples from hydrothermal vent systems in the southern Gulf of California, including Alarcón Rise (AR), Pescadero Basin (PB), and Pescadero Transform Fault (PTF)

	Observed OTUs	Faith PD	Shannon
PTF_2	2262	89.9	10
PTF_4	1936	84.6	9.3
PB_1	1981	79.5	9.3
PB_3	1221	61.9	7.6
PB_8	646	45.9	6.6
AR_5	1453	68.8	8.3

UPGMA Jaccard

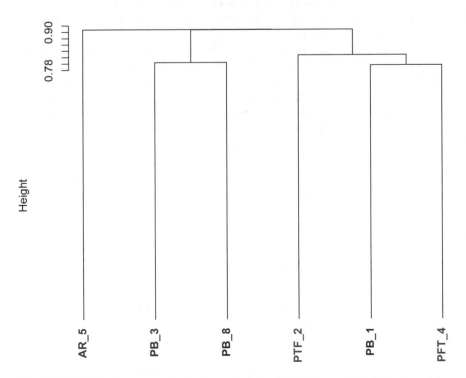

Fig. 7.3 UPGMA cluster analysis based on the beta diversity Jaccard dissimilarity matrix. Samples included six bacterial communities (Alarcón Rise (AR), Pescadero Basin (PB), and Pescadero Transform Fault (PTF)) from hydrothermal vent systems in the southern Gulf of California

rare biosphere (Sogin et al. 2006). This sizeable genetic pool contributes importantly to the metabolic potential of the community (Jousset et al. 2017) and is known to increase in response to environmental changes. It is possible that in the analyzed vent systems this rare biosphere plays a vital role in assuring the ecosystem functions, acting as ecologically important keystone organisms, as they outnumber abundant taxa (Kurm et al. 2019).

7.3.3 Microbial Ecological Roles (Metabolic Signatures)

Since abundant and shared species represent an important component of microbial communities, the shared (occurring in more than in a single site), abundant (≥1% reads), and "iconic" taxa (based on their reported metabolic capacities) within the OTU database were explored. Firstly, the largest number of shared OTUs among

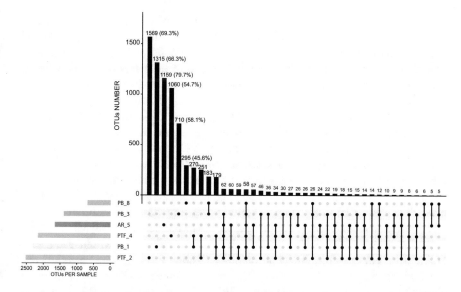

Fig. 7.4 Graphic representation of shared OTUs within the studied deep hydrothermal vent systems sites in the southern Gulf of California. Number of unique or shared OTUs (among samples) is represented by black bars; for unique OTUs, a percentage value is shown and was calculated with respect to the number of OTUs per site. Dots below bars denote the occurrence of particular OTUs in a specific site, and lines indicate sites shared OTUs between sites. The number of OTUs on each site is proportional to colored bars in the left of the main figure

sites, 179–270 (Fig. 7.4), reflected the clustering patterns as displayed by the beta diversity analysis (Fig. 7.3), where PB_1, PTF_2, and PTF_4 diverged from PB_3 and PB_8 and formed another cluster. Moreover, the low number of shared OTUs among all vent samples (58, 0.75% of the overall OTUs) highlighted again the uniqueness of each microbial community.

The exploration of abundant OTUs in the samples (Table 7.1) allowed the inference on microbial "metabolic signatures" that could reveal the potential effect of micro-niches driving the differentiation between microbial vent communities (Dick 2019), even in sites within the same system, since they are remarkably heterogeneous and exhibit sharp gradients. These "metabolic signatures" can reflect potential patterns for metabolic pathways associated with particular conditions, for example, the presence and concentration of chemical compounds, such as sulfur compounds, methane, petroleum, and other related hydrocarbons.

In hydrothermal vents, sulfur is a crucial element in sustaining microbial communities (Dick 2019). The oxidation of sulfur compounds has been extensively studied due to its importance in chemosynthetic pathways in hydrothermal fields (Sievert et al. 2008); on the other hand, less is known about sulfur compounds reduction (Cao et al. 2014). We found abundant hydrothermal vent sulfur/sulfide oxidizing bacteria (SOB), such as *Sulfurovum* (Mori et al. 2018) in PB and PTF samples (Table 7.1). These Epsilonbacteraeota are characteristic of sulfide-rich vent fluids (Akerman et al. 2013), and basalt-hosted vent sites of 9°N East Pacific Rise

(Campbell et al. 2013). Moreover, the contribution of this taxon to the nitrogen cycle has been reported as well (Li et al. 2018), highlighting its metabolic potential.

Furthermore, in AR we detected the sulfur compounds oxidizer abundant taxon *Thioalkalivibrio*. These Gammaproteobacteria have been reported within hydrothermal chimneys on the Southwest Indian Ridge (Cao et al. 2014). With respect to sulfur/sulfate reducing bacteria (SRB), we identified several Desulfobacterales OTUs in all our samples (Table 7.1), including unique OTUs (data not shown). This result agrees with previous work in Guaymas Basin, and the East Pacific Rise vent systems (Teske et al. 2002; Dhillon et al. 2003; Campbell et al. 2013); particular SRB taxa within Desulfobacterales, *Desulfobulbus* genus, were detected abundantly for AR, and Syntrophobacteraceae family showed more abundance for PTF_4.

The presence of sulfur/sulfide-oxidizing and sulfur/sulfate-reducing bacterial taxa (SOB and SRB) in all samples confirms the importance of sulfur compounds on the maintenance of vent communities, but the differences could reveal the "metabolic footprint" within each sample, such as particular taxa within each guild (Table 7.1). For example, AR is presenting a distinct sulfur-related bacterial community including *Thioalkalivibrio* (SOB) and *Desulfobulbus* (SRB). Another example is the lower levels of both SOB and SRB sulfur-related taxa in PB_1, pointing that this site could have different environmental characteristics than the other two PB samples. A third example is PB_4 showing high levels of SRB Syntrophobacteraceae taxa, in contrast with low levels of SOB taxa; the lack of observed sulfur compounds oxidizing prokaryotes leads us to hypothesize that methane, and not sulfur compounds, in PB_4 site could be the essential source of energy, based on the abundance of Methyloceanibacter, a methylotrophic taxon (Takeuchi et al. 2014) in this sample (Table 7.1). Interestingly Goffredi et al. (2017) reported the presence of methane in the end-member fluids of this area. Anaerobic oxidation of methane provides a notable source of energy and biomass at hydrothermal vents, in addition to the better-known oxidation of sulfur compounds by bacteria (Teske et al. 2002), and this metabolism could be coupled to the SRB (Muyzer and Stams 2008; Cassarini 2017). A further possibility is that SO archaea (Sievert et al. 2007; Nakagawa and Takai 2008) are filling this niche.

Sulfur and methane-related bacteria reflect the similarity among microbial taxa of PTF samples and PB_1, where we found Syntrophobacteraceae and Sva0081 taxa (both SRB taxa; Kuever et al. 2005; Campbell et al. 2013), and Methyloceanibacter (methane-related) in contrast to PB samples (PB_3 and PB_8) that were characterized by *Sulfurovum* microorganisms (SOB), or Class JS1 from Atribacteria, that are known to generate fermentation products that support methanogens (Carr et al. 2015). Since the PB_1 is the only sample within PB hydrothermal system that is located easterly (Fig. 7.1), one may expect a low influence on the sediments from venting fluids within this particular spot, triggering the observed lower abundance of SOB and SRB taxa (Zinke et al. 2018).

Epsilonbacteriaceae (SOB) abundance within the PB_3 and PB_8 samples (Fig. 7.2) coincides with the high levels of hydrogen reported for these sites (Goffredi et al. 2017; Paduan et al. 2018), given the ability of the taxa to use hydrogen for energy besides sulfur (Campbell et al. 2013). The abundance of these taxa in PTF_2 sample (Fig. 7.2) could point to the presence of hydrogen in this site. Whereas the reductive tricarboxylic acid cycle (rTCA), characteristic of Epsilonproteobacteria (Hügler et al. 2005; Nakagawa and Takai 2008; Cerqueira et al. 2015), could represent another "metabolic footprint" for PTF_2, PTF_4, and PB_1, equally the Calvin–Benson–Bassham cycle for PB_2, PB_8, and AR, due to the prevalence of Gammaproteobacteria (Hügler and Sievert 2011; Fig. 7.2).

The abundance of methane-related taxa (Table 7.1) demonstrated methane traces in PFT samples (Salcedo et al. 2019), whereas the lowest abundance of these taxa at AR confirmed the low concentrations of methane (Goffredi et al. 2017). Similarly, the occurrence of OTUs related with the degradation of long fatty acids in PB, including the MSBL8 family from Cloacimonetes, and *Thermovirga*, a genus isolated from oil sediment samples; (Dahle and Birkeland 2006; Shakeri Yekta et al. 2019), agrees with reports of long fatty chains in these vents (Goffredi et al. 2017).

The AR basalt-hosted system presented ferric metabolism-related taxa (Table 7.2) such as *Deferrisoma*, typically involved in iron reduction (Slobodkina et al. 2012), Acidiferrobacteraceae and *Mariprofundus*, regarded as Fe-oxidators (Emerson et al. 2007; Issotta et al. 2018); the occurrence of these bacterial taxa reflects the basalt high ferric concentration, which is characteristic of AR (Goffredi et al. 2017).

7.3.4 Deep Hydrothermal Systems as an Approximation to Understand Primeval Environments

To understand aqueous environments likely to have been present on the prebiotic Earth, and to investigate whether they would be conductive to the origin of life, we can learn from the different prebiotic analog sites that exist today in our planet, in particular from the submarine hydrothermal vents (Deamer et al. 2019). Different metabolic signatures found in this study between analyzed microbial communities from southern Gulf of California hydrothermal vent systems give consideration to diverse metabolic possibilities that reflect primordial life forms potentially originating in early Earth. Further detailed studies for these near (<100 km), but contrasting environmental systems, should be done, as a frontier for fruitful exploration, discovery, and research within this newly discovered Mexican ecosystem.

It has been proposed that in the Hadean (~ 4.6 to 4 billion years ago), in the absence of oxygen, alkaline vents acted as electrochemical flow reactors, in which alkaline fluids saturated in H_2 mixed with relatively acidic ocean waters rich in CO_2, through a labyrinth of interconnected micropores with thin inorganic walls containing catalytic Fe(Ni)S minerals (Sojo et al. 2016). The difference in pH across these thin barriers produced natural proton gradients with equivalent magnitude and polarity to the proton-motive force required for carbon fixation in present bacteria and archaea (Sojo et al. 2016).

Nowadays, alkaline vents form labyrinthine networks of interconnected micropores bounded by thin inorganic walls, through which hydrothermal fluids (and ocean waters) percolate. Such vents should have been even more common on the early Earth, as the mantle was less differentiated from the crust. Hence, ultramafic minerals could have been found across much of the ocean floor (Fyfe 1994; Shields and Kasting 2007; Jaffrés et al. 2007).

Along with the Cuatro Ciénegas Basin, where microbial communities with strong marine and hydrothermal vent resemblances (Souza et al. 2006) occur due to the presence of an active fault (Souza et al. 2018), these extreme Mexican systems (hydrothermal vents in the southern Gulf of California, and Cuatro Ciénegas hot springs) pose ideal models to explore primordial environments (e.g., the origin of life, early life, and astrobiology) where modern microorganisms metabolize ancient forms of chemical energy (de Anda et al. 2018).

7.4 Conclusions and Perspectives

Herein, we described the rich prokaryotic diversity in hydrothermal systems of the South of the Gulf of California represented by three different systems: PB, a carbonated high-temperature vent, AR dominated by basalt high-temperature vents, and PTF with low-temperature vents. All of which are characterized by particular sulfur compounds oxidizing and reducing bacteria, among other taxa. Remarkably, differentiated community structure assemblages among systems were observed, except for PB_1 clustering with PTF samples. This pattern may be elucidated by the detailed evaluation of the whole bacterial community.

Hydrothermal explorations pose future challenges and opportunities for science. The abundance of apparently primordial life-forms originating and still evolving in deep-sea vents and using rocky mineral walls in ocean-floor as power sources for chemical reactions opens up a world of possibilities to further explore a several hypotheses on the origin of life on Earth.

In this chapter, we provided evidence of the diverse chemotrophic microbial communities inhabiting high-temperature and low-temperature hydrothermal vents in the southern Gulf of California, and their characteristic "metabolic signatures" that are strongly related to local environmental conditions. The presence of these microbes confirms that life is possible in sea floor environments in hydrothermal systems, which were perhaps present and active during most of the early Earth.

Acknowledgments This work was supported by SEP- Ciencia Básica CONACyT Grant 238245 and Pappit-DGPA grant IG200319. We want to thank Dr. Erika Aguirre Planter for technical support. We are grateful to the Monterey Bay Aquarium Research Institute and especially to R. Vrijenhoek for his invaluable assistance in obtaining the samples herein analyzed, and for his unyielding support. A special word of appreciation to the Western Flyer crew and the ROV Doc Ricketts pilots for their invaluable help.

References

Akerman N, Butterfield D, Huber J (2013) Phylogenetic diversity and functional gene patterns of sulfur-oxidizing subseafloor Epsilonproteobacteria in diffuse hydrothermal vent fluids. Front Microbiol 4:185. https://doi.org/10.3389/fmicb.2013.00185

Álvarez-Borrego S (1983) Gulf of California. In: Ketchum, BH (ed) Estuaries and enclosed seas. Ecosystems of the world 26. Elsevier, Amstedama, pp 427–449

Baker ET, Lavelle JW, Feely RA et al (1989) Episodic venting of hydrothermal fluids from the Juan de Fuca ridge. J Geophys Res Solid Earth 94(B7):9237–9250

Barge LM, White LM (2017) Experimentally testing hydrothermal vent origin of life on Enceladus and other icy/ocean worlds. Astrobiology 17:820–833. https://doi.org/10.1089/ast.2016.1633

Baross JA, Hoffman SE (1985) Submarine hydrothermal vents and associated gradient environments as sites for the origin and evolution of life. Orig Life Evol Biosph 15:327–345

Beaulieu SE, Szafranski K (2018) InterRidge Global database of active submarine hydrothermal vent fields. Prepared for InterRidge, Version 34. http://vents-date.interridge.org

Bolyen E, Rideout JR, Dillon MR et al (2018) QIIME 2: reproducible, interactive, scalable, and extensible microbiome data science. PeerJ. https://doi.org/10.7287/peerj.preprints.27295v2

Bourbonnais A, Lehmann MF, Butterfield DA, Juniper SK (2012) Subseafloor nitrogen transformations in diffuse hydrothermal vent fluids of the Juan de Fuca Ridge evidenced by the isotopic composition of nitrate and ammonium. Geochem, Geophys Geosyst 13:Q02T01. https://doi.org/10.1029/2011GC003863

Bourbonnais A, Juniper SK, Butterfield DA et al (2014) Diversity and abundance of Bacteria and nirS-encoding denitrifiers associated with the Juan de Fuca Ridge hydrothermal system. Ann Microbiol 64:1691–1705. https://doi.org/10.1007/s13213-014-0813-3

Brazelton W (2017) Hydrothermal vents. Curr Biol 27:R450–R452

Campbell BJ, Polson SW, Allen LZ, et al (2013) Diffuse flow environments within basalt- and sediment-based hydrothermal vent ecosystems harbor specialized microbial communities. Front Microbiol 4. https://doi.org/10.3389/fmicb.2013.00182

Cao H, Wang Y, Lee OO, et al (2014) Microbial sulfur cycle in two hydrothermal chimneys on the Southwest Indian Ridge. mBio 5. https://doi.org/10.1128/mBio.00980-13

Carr SA, Orcutt BN, Mandernack KW, Spear JR (2015) Abundant Atribacteria in deep marine sediment from the Adélie Basin, Antarctica. Front Microbiol 6:872. https://doi.org/10.3389/fmicb.2015.00872

Cassarini C (2017) Anaerobic oxidation of methane coupled to the reduction of different sulfur compounds as electron acceptors in bioreactors. Dissertation, Université Paris-Est

Cerqueira T, Pinho D, Egas C et al (2015) Microbial diversity in deep-sea sediments from the Menez Gwen hydrothermal vent system of the Mid-Atlantic Ridge. Mar Genomics 24:343–355. https://doi.org/10.1016/j.margen.2015.09.001

Clague DA, Caress DW, Dreyer BM et al (2018) Geology of the Alarcon Rise, Southern Gulf of California. Geochem Geophys Geosyst 19:807–837. https://doi.org/10.1002/2017GC007348

Conway JR, Lex A, Gehlenborg N (2017) UpSetR: an R package for the visualization of intersecting sets and their properties. Bioinformatics 33:2938–2940. https://doi.org/10.1093/bioinformatics/btx364

Corliss JB, Dymond J, Gordon LI, et al (1979) Submarine thermal springs on the Galápagos Rift. Science 203:1073–1083

Corliss JB, Baross Ja, Hoffman S (1981) An hypothesis concerning the relationships between submarine hot springs and the origin of life on Earth. Oceanol Acta, Special issue 59–69

Dahle H, Birkeland NK (2006) *Thermovirga lienii* gen. nov., sp. nov., a novel moderately thermophilic, anaerobic, amino-acid-degrading bacterium isolated from a North Sea oil well. Int J Syst Evol Microbiol 56:1539–1545. https://doi.org/10.1099/ijs.0.63894-0

de Anda V, Zapata-Peñasco I, Eguiarte LE et al (2018) The sulfur cycle as the gear of the "clock of life": the point of convergence between geological and genomic data in the Cuatro Cienegas Basin. In: García-Oliva F, Elser J, Souza V (eds) Ecosystem ecology and geochemistry of Cuatro Ciénegas. Cuatro Ciénegas Basin: an endangered hyperdiverse Oasis. Springer, Cham, pp 67–83

de Wit MJ (1997) Early Archean processes: evidence from the South African Kaapvaal craton and its greenstone belts (extended abstract). Geol Mijnb 76:369–373. https://doi.org/10.102 3/A:1003290014097

Deamer D, Damer B (2017) Can life begin on Enceladus? A perspective from hydrothermal chemistry. Astrobiology 17:834–839. https://doi.org/10.1089/ast.2016.1610

Deamer D, Damer B, Kompanichenko V (2019) Hydrothermal chemistry and the origin of cellular life. Astrobiology 19:1523–1537. https://doi.org/10.1089/ast.2018.1979

Dhillon A, Teske A, Dillon J et al (2003) Molecular characterization of sulfate-reducing bacteria in the Guaymas basin. Appl Environ Microbiol 69:2765–2772. https://doi.org/10.1128/AEM.69.5.2765-2772.2003

Dick GJ (2019) The microbiomes of deep-sea hydrothermal vents: distributed globally, shaped locally. Nat Rev Microbiol 17:271–283

Dombrowski N, Teske AP, Baker BJ (2018) Expansive microbial metabolic versatility and biodiversity in dynamic Guaymas Basin hydrothermal sediments. Nat Commun 9: 4999. https://doi.org/10.1038/s41467-018-07418-0

Emerson D, Rentz JA, Lilburn TG, et al (2007) A novel lineage of proteobacteria involved in formation of marine Fe-oxidizing microbial mat communities. PLoS ONE 2:e667. https://doi.org/10.1371/journal.pone.0000667

Fyfe WS (1994) The water inventory of the Earth: fluids and tectonics. Geol Soc Lond Spec Publ 78:1. https://doi.org/10.1144/GSL.SP.1994.078.01.02

Ghai R, Mizuno CM, Picazo A et al (2013) Metagenomics uncovers a new group of low GC and ultra-small marine Actinobacteria. Sci Rep 3: 2471. https://doi.org/10.1038/srep02471

Goffredi SK, Johnson S, Tunnicliffe V et al (2017) Hydrothermal vent fields discovered in the southern gulf of California clarify role of habitat in augmenting regional diversity. Proc R Soc B Biol Sci 284:1–10. https://doi.org/10.1098/rspb.2017.0817

He T, Zhang X (2016) Characterization of bacterial communities in deep-sea hydrothermal vents from three oceanic regions. Mar Biotechnol 18:232–241. https://doi.org/10.1007/s10126-015-9683-3

Herschy B, Whicher A, Camprubi E et al (2014) An origin-of-life reactor to simulate alkaline hydrothermal vents. J Mol Evol 79:213–227. https://doi.org/10.1007/s00239-014-9658-4

Hessler RR, Kaharl VA (1995) The deep-sea hydrothermal vent community: an overview. Geophysical monograph series 91: 72–84

Hsu H-W, Postberg F, Sekine Y et al (2015) Ongoing hydrothermal activities within Enceladus. Nature 519:207–210. https://doi.org/10.1038/nature14262

Hügler M, Sievert SM (2011) Beyond the Calvin Cycle: autotrophic carbon fixation in the ocean. Annu Rev Mar Sci 3:261–289. https://doi.org/10.1146/annurev-marine-120709-142712

Hügler M, Wirsen CO, Fuchs G et al (2005) Evidence for autotrophic CO_2; fixation via the Reductive Tricarboxylic Acid Cycle by members of the ε subdivision of Proteobacteria. J Bacteriol 187:3020. https://doi.org/10.1128/JB.187.9.3020-3027.2005

Issotta F, Moya-Beltrán A, Mena C et al (2018) Insights into the biology of acidophilic members of the Acidiferrobacteraceae family derived from comparative genomic analyses. Res Microbiol 169:608–617. https://doi.org/10.1016/j.resmic.2018.08.001

Jaffrés JBD, Shields GA, Wallmann K (2007) The oxygen isotope evolution of seawater: a critical review of a long-standing controversy and an improved geological water cycle model for the past 3.4 billion years. Earth Sci Rev 83:83–122. https://doi.org/10.1016/J.EARSCIREV.2007.04.002

Jousset A, Bienhold C, Chatzinotas A et al (2017) Where less may be more: how the rare biosphere pulls ecosystems strings. ISME J 11:853–862

Kelley DS, Karson JA, Blackman DK et al (2001) An off-axis hydrothermal vent field near the Mid-Atlantic Ridge at 30 degrees N. Nature 412:145–149

Kelley DS, Karson JA, Früh-Green GL et al (2005) A serpentinite-hosted ecosystem: the Lost City hydrothermal field. Science 307:1428–1434. https://doi.org/10.1126/science.1102556

Kuever J, Rainey FA, Widdel F (2005) Order VI. Syntrophobacterales ord. nov. In: Brenner DJ, Krieg NR, Staley JT, Garrity GM (eds) Bergey's manual of systematic bacteriology, the Proteobacteria (the Aplha-, Beta-, Delta-, and Epsilonproteobacteria), vol 2. Springer, New York, p 1021

Kurm V, Geisen S, Gera Hol WH (2019) A low proportion of rare bacterial taxa responds to abiotic changes compared with dominant taxa. Environ Microbiol 21:750–758. https://doi.org/10.1111/1462-2920.14492

Li Y, Tang K, Zhang L, et al (2018) Coupled Carbon, Sulfur, and Nitrogen cycles mediated by microorganisms in the water column of a shallow-water hydrothermal ecosystem. Front Microbiol 9. https://doi.org/10.3389/fmicb.2018.02718

Martin W, Baross J, Kelley D, Russell MJ (2008) Hydrothermal vents and the origin of life. Nat Rev Microbiol 6:805–814. https://doi.org/10.1038/nrmicro1991

Mori K, Yamaguchi K, Hanada S (2018) *Sulfurovum denitrificans* sp. Nov., an obligately chemolithoautotrophic sulfur-oxidizing epsilonproteobacterium isolated from a hydrothermal field. Int J Syst Evol Microbiol 68:2183–2187. https://doi.org/10.1099/ijsem.0.002803

Muyzer G, Stams AJM (2008) The ecology and biotechnology of sulphate-reducing bacteria. Nat Rev Microbiol 6:441–454. https://doi.org/10.1038/nrmicro1892

Nakagawa S, Takai K (2008) Deep-sea vent chemoautotrophs: diversity, biochemistry and ecological significance. FEMS Microbiol Ecol 65:1–14

Oksanen J, Blanchet FG, Kindt R, et al (2013) Vegan: community ecology package. R package version 2.0-8. http://CRAN.R-project.org/package=vegan

Paduan JB, Zierenberg RA, Clague DA et al (2018) Discovery of hydrothermal vent fields on Alarcón Rise and in Southern Pescadero Basin, Gulf of California. Geochem Geophys Geosyst 19:4788–4819. https://doi.org/10.1029/2018GC007771

Pjevac P, Meier DV, Markert S et al (2018) Metaproteogenomic profiling of microbial communities colonizing actively venting hydrothermal chimneys. Front Microbiol 9. https://doi.org/10.3389/fmicb.2018.00680

Rudi K, Skulberg OM, Larsen F, Jakobsen KS (1997) Strain characterization and classification of oxyphotobacteria in clone cultures on the basis of 16S rRNA sequences from the variable regions V6, V7, and V8. Appl Environ Microbiol 63:2593–2599

Russell MJ, Hall AJ (1997) The emergence of life from iron monosulphide bubbles at a submarine hydrothermal redox and pH front. J Geol Soc 154:377–402

Russell MJ, Daniel RM, Hall AJ (1993) On the emergence of life via catalytic iron-sulphide membranes. Terra Nova 5:343–347. https://doi.org/10.1111/j.1365-3121.1993.tb00267.x

Salcedo DL, Soto LA, Paduan JB (2019) Trophic structure of the macrofauna associated to deep-vents of the southern Gulf of California: Pescadero Basin and Pescadero transform fault. PLoS ONE 14: e0224698. https://doi.org/10.1371/journal.pone.0224698

Shakeri Yekta S, Liu T, Axelsson Bjerg M et al (2019) Sulfide level in municipal sludge digesters affects microbial community response to long-chain fatty acid loads. Biotechnol Biofuels 12:259. https://doi.org/10.1186/s13068-019-1598-1

Shields GA, Kasting JF (2007) Palaeoclimatology: evidence for hot early oceans? Nature 447: E1

Sievert SM, Wieringa EBA, Wirsen CO, Taylor CD (2007) Growth and mechanism of filamentous-sulfur formation by Candidatus Arcobacter sulfidicus in opposing oxygen-sulfide gradients. Environ Microbiol 9:271–276. https://doi.org/10.1111/j.1462-2920.2006.01156.x

Sievert SM, Hügler M, Taylor CD, Wirsen CO (2008) Sulfur oxidation at deep-sea hydrothermal vents. In: Dahl C, Friedrich CG (eds) Microbial sulfur metabolism. Springer, Berlin Heidelberg, Berlin, Heidelberg, pp 238–258

Slobodkina GB, Reysenbach AL, Panteleeva AN et al (2012) Deferrisoma camini gen. nov., sp. nov., a moderately thermophilic, dissimilatory iron (III)-reducing bacterium from a deep-sea hydrothermal vent that forms a distinct phylogenetic branch in the Deltaproteobacteria. Int J Syst Evol Microbiol 62:2463–2468. https://doi.org/10.1099/ijs.0.038372-0

Sogin ML, Morrison HG, Huber JA et al (2006) Microbial diversity in the deep sea and the underexplored "rare biosphere". Proc Natl Acad Sci U S A 103:12115–12120. https://doi.org/10.1073/pnas.0605127103

Sojo V, Herschy B, Whicher A et al (2016) The origin of life in alkaline hydrothermal vents. Astrobiology 16:181–197. https://doi.org/10.1089/ast.2015.1406

Souza V, Espinosa-Asuar L, Escalante AE et al (2006) An endangered oasis of aquatic microbial biodiversity in the Chihuahuan desert. Proc Natl Acad Sci U S A 103:6565–6570. https://doi.org/10.1073/pnas.0601434103

Souza V, Eguiarte LE, Elser JJ, et al (2018) A microbial saga: how to study an unexpected hot spot of microbial biodiversity from scratch?. In: Souza V, Olmedo-Álvarez G, Eguiarte L (eds) Cuatro Ciénegas ecology, natural history and microbiology. Cuatro Ciénegas Basin: an endangered hyperdiverse oasis. Springer, Cham, pp 1–20

Takeuchi M, Katayama T, Yamagishi T et al (2014) Methyloceanibacter caenitepidi gen. nov., sp. nov., a facultatively methylotrophic bacterium isolated from marine sediments near a hydrothermal vent. Int J Syst Evol Microbiol 64:462–468. https://doi.org/10.1099/ijs.0.053397-0

Teske A, Hinrichs KU, Edgcomb V et al (2002) Microbial diversity of hydrothermal sediments in the Guaymas Basin: evidence for anaerobic methanotrophic communities. Appl Environ Microbiol 68:1994–2007. https://doi.org/10.1128/AEM.68.4.1994-2007

Troussellier M, Escalas A, Bouvier T, Mouillot D (2017) Sustaining rare marine microorganisms: macroorganisms as repositories and dispersal agents of microbial diversity. Front Microbiol 8:947. https://doi.org/10.3389/fmicb.2017.00947

Von Damm KL (2001) Lost city found. Nature 412(6843):127

Wächtershäuser G, User C (1992) Growndworks for an evolutionary biochemistry: the iron-sulphur world. Prog Biophys Mol Biol 58: 85–201

Waite DW, Vanwonterghem I, Rinke C et al (2017) Comparative genomic analysis of the class Epsilonproteobacteria and proposed reclassification to Epsilonbacteraeota (phyl. nov.). Front Microbiol 8:682. https://doi.org/10.3389/fmicb.2017.00682

Zhang L, Kang M, Xu J et al (2016) Bacterial and archaeal communities in the deep-sea sediments of inactive hydrothermal vents in the Southwest India Ridge. Sci Rep 6: 25982. https://doi.org/10.1038/srep25982

Zierenberg RA, Adams MW, Arp AJ (2000) Life in extreme environments: hydrothermal vents. Proc Natl Acad Sci U S A 97:12961–12962. https://doi.org/10.1073/pnas.210395997

Zinke LA, Reese BK, McManus J, et al (2018) Sediment microbial communities influenced by cool hydrothermal fluid migration. Front Microbiol 9:1249. https://doi.org/10.3389/fmicb.2018.01249

Chapter 8
Andean Microbial Ecosystems: Traces in Hypersaline Lakes About Life Origin

Luis A. Saona, Mariana Soria, Patricio G. Villafañe, Agustina I. Lencina, Tatiana Stepanenko, and María E. Farías

Abstract High-altitude Andean lakes (HAALs) represent unique environments on the Earth where one can study the biological chemistry of life in one of its most extreme versions. The Atacama Desert, Argentine Puna, and Bolivian Altiplano harbor hypersaline lakes where polyextremophilic Andean Microbial Ecosystems (AMEs) inhabit microbial mats, evaporitic mats, biofilms (BF), evaporites (EV), and microbialites (Mi). These AMEs have two remarkable characteristics: (i) they are the only ones in the world that inhabit areas ranging from 3100 to 4200 masl; and (ii) they are excellent modern analogues of those which populated the primitive Earth ~3 billion years ago. In this chapter, we will delve into the different kinds of AMEs present in the HAAL, their formation, structure, and their adaptation to conditions largely influenced by volcanic activity, UV radiation, arsenic content, high salinity, low dissolved oxygen content, extreme daily temperature fluctuation, and oligotrophic conditions. All of these physicochemical parameters recreate the early Earth and even extraterrestrial conditions. The relevance of studying these ecosystems does not lie only in scientific-descriptive and/or economic interest. The scientific research community has a great responsibility to address climate change. In this scenario, the AMEs could have played a key role, influencing changes that allowed the origin of aerobic life and those who have faced the great climatic events of the Earth.

8.1 Introduction

The hypersaline environments, ponds and lakes, are mainly dominated by NaCl (up to saturation concentrations), and they are alkaline ecosystems (Risacher et al. 2003). They are widely distributed across the globe but are highly abundant in

L. A. Saona · M. Soria · P. G. Villafañe · A. I. Lencina · T. Stepanenko · M. E. Farías (✉)
Laboratorio de Investigaciones Microbiológicas de Lagunas Andinas (LIMLA), Planta Piloto de Procesos Industriales Microbiológicos (PROIMI), CCT, CONICET,
San Miguel de Tucumán, Argentina

© Springer Nature Switzerland AG 2020
V. Souza et al. (eds.), *Astrobiology and Cuatro Ciénegas Basin as an Analog of Early Earth*, Cuatro Ciénegas Basin: An Endangered Hyperdiverse Oasis,
https://doi.org/10.1007/978-3-030-46087-7_8

167

terrestrial biomes, such as deserts and steppes and in geologically interesting regions including the East African Rift Valley. Other geographically well-known hypersaline environments are Great Salt Lake (Utah), the Dead Sea (Israel), and the alkaline lakes of Wadi Natrun (Egypt) (Larsen 1980).

Environments, such as Puna (Argentina), Altiplano (Bolivia), or Atacama Desert (Chile), exhibit unique geography, and they are a large reservoir of high-altitude Andean lakes (HAAL). These lakes have wide environmental factors, such as UV radiation, arsenic content, high salinity, low dissolved oxygen content, extreme daily temperature fluctuation, and oligotrophic conditions, that represent a window to the early Earth and even extraterrestrial conditions where currently we are looking for life, like Mars (Farías et al. 2014; Albarracín et al. 2015).

Despite the extreme conditions, various substrates in these environments are colonized by rich and diverse communities of phototrophic and heterotrophic microorganisms, which are referred to as Andean Microbial Ecosystems (AME). The AMEs are complex associations of microbes that in many cases influence or induce mineral precipitation in lakes, hydrothermal vents, puquios, fumaroles of volcanoes, and salt flats of the Central Andes Region (Southwest Bolivia, Northern Argentina, and Chile) (Gomez et al. 2014). These ecosystems include:

- Biofilms, which are communities of microbial cells associated with a solid surface and embedded in an exopolymeric substance (EPS) matrix (Marshall 1984).
- Microbial mats, which are benthic communities growing on a solid substrate (e.g., sand, rock, or sediments) as vertical layers of functional groups of microbes embedded in an mineral-organic matrix, more often exopolysaccharides, silicates, and carbonates (Dupraz et al. 2009; Glunk et al. 2011; Albarracín et al. 2016a).
- Microbialites, which are microbial mats that present various degrees of mineral precipitation and lithification (Dupraz et al. 2009; Glunk et al. 2011; Albarracín et al. 2016a).
- Endoevaporitic domes, which are microbial mats associated with gypsum crusts and halite.

Some of the Puna-Altiplano microbialites resemble those observed in the ancient geological record, for example, stromatolites, which are the oldest evidence of life on Earth (Allwood et al. 2006; Nutman et al. 2016, 2019). Thus, these systems, inhabited by microorganisms known as "polyextremophiles,," are excellent microbial ecology models, and they provide us a unique opportunity to apply an integrated geobiological approach to study the biochemist and molecular mechanisms of the adaptation to extreme conditions like those that existed 3800 million years ago.

This chapter describes the HAALs, as a relatively unexplored natural laboratory, where biofilms, microbial mats, endoevaporitic domes, and microbialites can be found. These HAALs can be considered the modern counterparts of the ancient microbial ecosystems. Analyzing all the physicochemical parameters that make the arid Central Andean endorheic basins placed on Argentina (Puna), Chile (Atacama Desert), and Bolivia (Altiplane) as one of the most extreme places on Earth. Lastly, we will focus on the molecular and biochemist tools of AMEs to resist all these abiotic stress and even use it. All this information gives us interesting insights into the ancient microbial life.

8.2 Life's Origin

A long time ago, specifically in the Precambrian ocean, life arose (Mojzsis et al. 1996; Rosing 1999; Watanabe et al. 2000; Kempe and Kazmierczak 2002), and from there it began its evolutionary path to the cells that inhabited different environments. The dating of these events is not entirely clear, and until a few years ago, the consensus was that the oldest life records found to date corresponded to organo-sedimentary structures, known as stromatolites, located in the Pilbara Block of Western Australia dating back to 3.48 billion years back (Walter et al. 1980; Van Kranendonk et al. 2008). These fossils and their dating suggested that the origin of life occurred a few hundred million years before, between 3.7 and 3.8 billion years ago (Watanabe et al. 2000; Kempe and Kazmierczak 2002). However, stromatolites reported in the Southwest of Greenland in Isua Supracrustal Belt date back 3.7 billion years ago (Nutman et al. 2016, 2019) moving the date of the origin of life beyond 4000 million years.

Four thousand million years ago, the Earth was in the period of time called Archean; this eon is characterized by being between the first geological evidence and the next 2500 million years. There, our last unique common ancestor (LUCA) evolved and lived (Fox et al. 1980; Weiss et al. 2016), on a planet that, despite the huge differences with today's Earth was already habitable (Benedetto 2010).

The primitive Earth did not have the same physical-chemical composition as today. The atmosphere of that time was formed gradually by the degasification of the mantle and the gases contributed by the Great Meteorite Bombings. Although its original composition is still a matter of speculation, there is a consensus on it. In the atmospheric composition, gases, such as H_2O, CO, CO_2, H_2, HCN, NH_3, and CH_4 initially predominated; and later the volcanic activity would have contributed gases like N_2, CO_2, and H_2O (Pla-García and Menor-Salván 1990; Holland 1994). The first organisms to inhabit the Earth were prokaryotic microorganisms that carried out anoxygenic photosynthesis, which is characterized for not releasing dioxygen as a waste product (Kulp et al. 2008; Shih 2015). Later, cyanobacteria evolved and developed oxygenic photosynthesis with which they began to release oxygen into the atmosphere (Holland 1997; Summons et al. 1999; Kasting 2001). In this way, it was not until 2.5 billion years ago when the level of oxygen in the atmosphere began to increase in the event known as great oxygenation event (GOE) (Shih 2015). This event allowed to anaerobic life began to breathe oxygen and could evolve giving way to the great diversity of aerobic organisms that inhabit the Earth today.

In this metabolic context, the role of other elements that could have been a key for the development of life has been widely studied. In this chapter, we will focus on the role of the element arsenic. During the Achaean, the highest concentrations of As were found in rock bodies called "green rock belts" (i.e., metamorphized into green-amphibolite shale facies and intruded by granite bodies), because they contain important massive sulfide deposits, thereby concentrating the As (Lahaye et al. 2001; Bundschuh et al. 2008; Benedetto 2010).

8.3 High-Altitude Andean Lakes as Analogues to Primitive Earth

8.3.1 Physicochemical Conditions in Andean Valleys

We already mentioned the physicochemical characteristics of the primitive Earth where life originated; now let's think about how that land looked like. Many images exist that try to represent the landscapes of the ancient Earth of 4000 million years ago; however, these are usually characterized by a mountainous cordon in the horizon with reddish-brown colors and its active volcanoes. The water that joins that horizon with the bay at the bottom of the image where the first life forms of which there is register are outlined: stromatolites. It is possible to find very similar primitive landscapes in the valleys of the Central Andes (Fig. 8.1).

The Central Andes region is home to a set of physicochemical, geological, and mineralogical conditions very similar to those that are presumed to have existed on the primitive Earth where life originated. This high-altitude region that encompasses Argentine, Chilean, and Bolivian territory is characterized by closed basins where salt flats, volcanoes, hypersaline lakes, hydrothermal fountains, and deserts abound. Due to its southern location and its altitude gradients, it is subjected to a set of environmental conditions that transform it into one of the most extreme places on the planet: low oxygen pressure, high UV radiation, constant volcanic eruptions, high thermal fluctuation, and high aridity, among others (Vuille et al. 2000, 2004; Garreaud et al. 2009; Albarracín et al. 2015). At this point, we will place special emphasis on UV radiation, dryness, salinity, and the presence of toxic elements (metals) for most life forms.

In terms of UV radiation, the Andean Valleys are one of the most irradiated areas on the planet. The monthly average reaches 6.6 kW h m^{-2} d^{-1} with maximum irradiation of 310 W m^{-2} (Duffie and Beckman 2013). These high levels of radiation are due in part to the latitude, which offers small midday solar zenith angles throughout the year, in addition to clean atmospheres at an average height of 3600 m above sea level (masl) with low aerosol content and low amount of clouds that reduces the attenuation of solar radiation (Luccini et al. 2006).

This high radiation consequently affects the aridity of the region. Although from January to April maximum rainfall is received, in a phenomenon called Bolivian winter, there is low humidity (38.5%) and low precipitation (1.47 mm) in this region (Albarracín et al. 2015). In the Atacama Desert, located from 19° S to 30° S in northern Chile and extending from the Pacific coast to the Andes mountain range (Davila et al. 2008; Contador et al. 2019), there are salt crusts present (halite and plaster) that can be considered the last niche available for life in hyperarid environ-

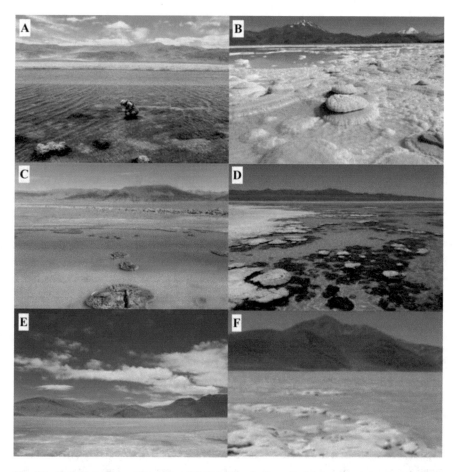

Fig. 8.1 Overview of sampling sites. (**A**) Sampling in Pozo Bravo Lake in Salar de Antofalla, Catamarca, Argentina. At the bottom of photo, the stromatolites are presents. In the photo samples are being taken of microbial mats inhabiting the ground of lake. (**B**) Endoevaporitic systems in Laguna 3 Quebradas, Catamarca, Argentina. (**C**) Microbialites present in Carachipampa, Catamarca, Argentina. (**D**) Laguna La Brava, Antofagasta, Chile. In the bottom of the photo, microbial mats can be observed. (**E**) Laguna Diamante, Argentina. This lake is the crater of Gálan volcano. (**F**) Stromatolites of Laguna Socompa, Argentina

ments on Earth (Connon et al. 2007; Davila et al. 2008; Schulze-Makuch et al. 2018). These aridity conditions in the Atacama Desert date back 150 million years ago, making it not only the driest on the planet but also one the oldest (Hartley et al. 2005; Clarke 2006).

Finally, due to the volcanic activity of the region (Smedley and Kinniburgh 2002), HAALs present high arsenic concentration (Demergasso et al. 2007; Escudero et al. 2013; Rascovan et al. 2016; Kurth et al. 2017). The average concentration varies between 0.8 mg L^{-1} and 11.8 mg L^{-1} (Albarracín et al. 2015). The highest As concentration was found in Diamante Lake, located in the crater of Galán

Volcano, with an As concentration of 230 mg L^{-1} (Rascovan et al. 2016), which are almost ten times higher than the concentration present in Mono Lake, a lake that has been studied for its high As concentration (Oremland et al. 2009; Wolfe-Simon et al. 2011). New measurements made in the water column of this lake, report up to 300 mg L^{-1}. This high concentration of As indisputably influences over the metabolism of the microorganisms that inhabit these lakes.

In summary, the Andean Valleys region has climatic, geological, atmospheric, and mineralogical conditions that make it the best analogue of the primitive Earth. All these abiotic characteristics directly influence the selection, adaptation, and evolution of the life that is a fundamental part of the landscapes of the Andes. In this sense, both the HAALs and the AMEs that inhabit them represent a natural laboratory from which we can study and find more clues about the origin of life. Although these ecosystems are modern and do not date from 3.7 million years ago, they are the same type of life – referred to as biofilm, microbial mats, and/or organo-sedimentary structures (microbialites) – that arose in the primitive Earth. In this sense, during the last decade, these ecosystems have been studied from metagenomic, genomic, physiological, and biogeochemical approaches; areas in which our research group has provided information specially referred to (i) abundance and kind of ecosystems as well as the (ii) biological role of arsenic as part a primitive metabolism for obtaining energy.

8.3.2 Andean Microbial Ecosystems (AMEs)

The Puna-Altiplano region has become an interesting region to study microbial mineralization processes. The most widely studied types of microbial-induced organomineralization are those of carbonate with calcium, as calcite and/or aragonite (Reid et al. 2000), calcium and magnesium (Vasconcelos and McKenzie 1997; Glunk et al. 2011), and magnesium (Thompson and Ferris 1990; Sanz-montero and Rodríguez-Aranda 2008).

Extreme changes in environmental factors, like salinity and temperature, together with a favorable local geology and geomorphology, facilitate the formation of closed lakes and groundwater springs, which are saturated in minerals, including carbonates. Thus, a highly mineralizing setting together with physicochemical and microbiological processes triggers carbonate precipitation and produces a set of diverse organo-sedimentary structures, referred as AMEs (Burne and Moore 1987).

As previously stated, the AMEs are complex associations of microbes that generally are dominated by Proteobacteria, Bacteroidetes, and Firmicutes. Another particularity of these systems is the presence of Cyanobacteria, which thrive as primary producers. Diatoms and fungi may also be present of these ecosystems (Farías et al. 2014; Rasuk et al. 2014; Fernandez et al. 2016).

Our research group worked in the analysis of the AMEs in the Puna-high Andes region. In 2009, the first microbialite was reported in Laguna Socompa, followed by Laguna Diamante in Argentina (Belluscio 2010) (Fig. 8.1). Since then, an extensive surveying has been carried out to the salt flats, lagoons, geysers, and fumaroles in the highlands of Bolivia, Argentina, and Chile. For 10 years, 80 environments were examined, and 20% of them presented at least 1 type of these microbial ecosystems. They include biofilms, microbial mats, microbialites, and endoevaporite inhabiting gypsum domes.

All throughout the examination of the AMEs, some shared characteristics were found, from a geological, physical, and chemical point of view: (i) active volcanic incidence – all the microbial ecosystems found where in some way connected to areas were active volcanoes are present. (ii) Underground water input. (iii) Mixed zones with different salinities: underground low conductivity water and salar thalassic water. (iv) Two main organo-sedimentary structures were found: oncolites (Laguna Negra, Tres Quebradas, and Las Quinoas, among others) and microbial mounds with a composite internal fabric of mesoclots and individual laminae. The mounds facies consist of thrombolites (clotted fabrics), which commonly grade upward and outward to stromatolites caps (laminae) (La Brava, Pozo Bravo, Ojo Bravo, El Peinado). From the biological point of view: (v) Predominance of diatoms, the main component in all the studied systems. (vi) Predominance of anaerobic over aerobic photosynthetic microorganisms. (vii) Microbial rhodopsin as the main system for producing ATP. (viii) Arsenic resistance and bioenergetic mechanisms. (ix) Predominance of carbon fixation pathways alternative to the Calvin cycle.

The knowledge and recent research activities in these systems have served as the basis for the declaration of protected areas in Laguna Socompa, Puquios de Mar, Tolar Grande, Llamara, and Tebenquiche. Additionally, AMEs are now included in baselines and environmental impact studies for lithium mining company projects in Chile's Salar de Atacama (Albemarle) and Catamarca, Argentina (Liex, Albemarle, Ultra Lithium and Morena del Valle Minerals). Taken together, the preservation of these unique ecosystems is a priority due to the benefits of AEMs to not only scientific understanding of microbial adaptation to extreme environments, but also these systems can be of great relevance to the economic development of local Andean populations.

8.3.3 Early Metabolisms in Andean Valleys

As we detailed above, due to the poly-extreme conditions present in the HAALs and in the Andean Valleys in general, the AMEs have selected specific genes that allow them to have highly specialized metabolisms.

In the case of extreme UV radiation, microbial communities have efficient repair mechanisms for accumulated DNA damage in the form of cyclobutane pyrimidine dimer (Fernández Zenoff et al. 2006; Albarracín et al. 2012); express different photolyases to minimize the damage of direct exposure to solar radiation (Sancar 2003; Albarracín et al. 2013); synthesize antioxidant enzymes to prevent damage from oxidative stress (Albarracín et al. 2011; Di Capua et al. 2011); produce secondary metabolites, such as pigments (Weber 2005; Kurth et al. 2015; Albarracín et al.

2016b); and form biofilms associated with minerals and rocks also to avoid direct exposure to UV rays (Albarracín et al. 2016a; Rascovan et al. 2016).

AMEs must also be able to access water molecules in one of the driest deserts in the world, which represents the last niche available for life in hyperarid environments (Davila et al. 2008). In this sense, the endolithic microbial communities of the Atacama Desert, composed mainly of Cyanobacteria, inhabit evaporitic rocks of sulfate and halite and appear to settle in the areas of maximum available humidity (Rothschild 1990; Parnell et al. 2004; Dong et al. 2007). In addition, due to the oxygen solubility decreases as a higher salt concentration, AMEs develop under microaerophilic and anoxic conditions.

In the primitive Earth, around 3.8 billion years ago, the high exposure to UV radiation, together with other conditions such as the alkaline pH of hydrothermal sources, favored the biogenesis of macromolecules, either allowing the accumulation of organic matter (avoiding acid hydrolysis) or contributing as source of energy for the development of complex molecules relevant to life (Kempe and Kazmierczak 2002; Rapf and Vaida 2016). Similarly, it is presumed that high salinity was an important factor in the formation of macromolecules from smaller organic molecules, due to the low activity of water in these conditions (Knauth 1998). In addition to these mechanisms, one of the metabolisms that we have deeply studied and that is related to the reactions that were presumably present in the origin of life is the arsenic metabolism.

Arsenic is a chemical element belonging to metalloids, although it is present along the Earth's crust, its relative abundance is very low and is considered an extremely toxic element. An interesting aspect of arsenic in relation to its role in biotic chemistry is that it is structurally similar to phosphorus; both have the same oxidation state, atomic radius, and electronegativity. In its most characteristic oxidized form (+5), arsenate and phosphate have a negative charge at physiological pH. In fact, both molecules have similar speciation at different pH, as well as similar pKa values (Anbar et al. 2009; Tawfik and Viola 2011) and thermochemical radii, which differ by only 4% (Kish and Viola 1999).

As a result of the high volcanic activity, the As concentration in HAALs is very high (Rascovan et al. 2016; Corenthal et al. 2016; Boutt et al. 2016). However, despite their high toxicity, AMEs have specific genes that allow them to survive under these conditions (Fig. 8.2). Here, we detail a brief summary of As biological role studied in different AMEs.

8.3.3.1 Laguna Diamante

Laguna Diamante is placed at 4560 masl inside Volcano Gálan caldera at Argentinean Puna (Fig. 8.1). This Lake presents extreme conditions of high conductivity (180 mS), alkalinity (pH 10), and As concentrations reach 230 mg L^{-1}. In this lake, a red biofilm associated to gaylussite mineral and volcanic rock is present. This microbial community is composed of more than 94% Haloarchaea. Previous metagenomic analyses of red biofilm samples have revealed a high abundance of

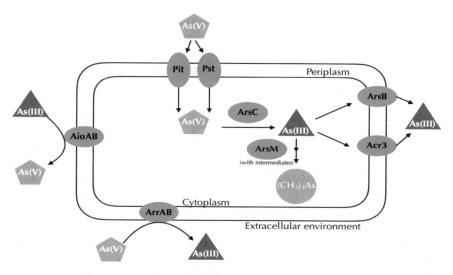

Fig. 8.2 Scheme of arsenic metabolic cycle

genes used for arsenite oxidation (e.g., *aioAB*) (Fig. 8.2) and respiratory arsenate reduction (e.g. *arrAB)* (Fig. 8.2) (Rascovan et al. 2016). The theorizing of this process has shown that cells can breathe As and thus obtain energy from their oxidation and reduction (Andres and Bertin 2016; Rascovan et al. 2016; Ordoñez et al. 2018). This specific point creates a direct link to prebiotic chemistry: electrochemical transfer through oxidation reactions by the *aio* gene is present before the divergence between bacteria and archaea and even from times before the last universal common ancestor (LUCA) (Lebrun et al. 2003; Kulp 2014). In other words, it is theorized that even before the formation of the first cell, reactions between arsenic species were already catalyzed by means of proto-enzymes.

Physiological assays with *Halorubrum* DM2 sp. (Laguna Diamante isolate) where arsenic facilitated its growth suggested an important role of arsenic as a bioenergetic component (Ordoñez et al. 2018). In this strain, an increment in the transcriptional levels of the genes related to arsenic metabolism when arsenic was added to the growth medium was observed by RT-qPCR (Stepanenko unpublished). Although it is not yet possible to assign the underlying molecular mechanism, these results suggest the idea that the biogeochemical cycle of arsenic is affected by microbial metabolism and vice versa.

8.3.3.2   Laguna La Brava

La Brava placed in Salar de Atacama, Chile, at 2200 masl (Fig. 8.1) is drained with groundwaters containing leached volcanic material (Corenthal et al. 2016; Boutt et al. 2016), in which high concentrations of arsenic and sulfide are found. Hence, the microbial metabolisms in the mats are possibly driven by anoxygenic photosynthesis using reduced sulfur and arsenic compounds (Visscher et al., in prep). Anaerobic processes could include fermentation, methanogenesis, sulfate, and likely arsenate reduction.

In regard to As, due to widespread volcanism and geothermal activity character- istic of the Precambrian, considerably more arsenic was accumulated on the Earth's surface than today (Cabral and Beaudoin 2007; Witt-Eickschen et al. 2009). Hence, as modern counterpart of Precambrian system, a living microbial mat from Laguna Brava has been investigated, using an array of conventional geochemical techniques (Farias et al. 2017; Sancho-Tomás et al. 2018). This experimental approach allowed to unravel the relationship between the microbial mat activity, mineral occurrence, arsenic speciation, distribution of major and trace elements, and their relationship with the mineralogy and the exopolymeric substances (EPS). In that way it was shown that As was not linked to Ca or Si, and only moderately related to Fe, result- ing from sorption onto an iron (oxy)hydroxide mineral. In that way, a number of features that support the biological cycling of arsenic have being documented (Sancho-Tomás et al. 2018). These include the incorporation of arsenic into exo- polymeric substances, the identification of organic-rich globules containing abun- dant arsenic with no other trace metals, and the heterogeneous distribution of As(III) and As(V), either together or separately and at different scales. Although accurate information on sulfur distribution and speciation are missing, the high concentra- tion of both sulfur and arsenic in the water column supports the notion of a coupled arsenic-sulfur cycle (Sancho-Tomás et al. 2018).

8.3.3.3 Laguna Socompa

Laguna Socompa is placed in Argentinean Puna at 3400 masl in the basis of the homonym volcano (Fig. 8.1). As concentration in Socompa lake is 28 mgL^{-1} whereas in the microbialite the salinity rises to 32 mg L^{-1} (Farías et al. 2013). The arsenic resistance mechanism developed by these microbial communities consisted mainly on the reduction and extrusion of As (V) by diverse Acr3 efflux pumps, associated to ArsC (thioredoxin type) reductases, which was also observed in strains that were isolated from Socompa (Kurth et al. 2017; Ordoñez et al. 2018). In the stromatolites, a complete set of arsenic metabolism genes was unraveled (Kurth et al. 2017), including *arsM* methyltransferases from the most varied phyla, as well as respiratory *aioA* oxidases, *arrA* reductases, and even novel *arxA* oxidases; there- fore, the community is able to carry out a complete arsenic cycle, even using it as a source of energy, allowing growth and adaptation of microorganisms to high con- centration of arsenic.

8.4 Conclusions

Andean Microbial Ecosystems (AMEs) represent a unique reservoir of microbial life adapted to extreme conditions; among them, arsenic resistance mechanisms and bioenergetic pathways have developed in efficient strategies that allow these communities to colonize all kinds of extreme ecosystems, from stromatolites developing in volcanic influenced areas like Socompa lake (3400 masl) to gaylussite biofilms evolving at 230 mg L^{-1} of As in an volcanic caldera at Volcano Gálan at 4560 masl. The association of arsenic geochemical cycle associated to anoxygenic photosynthesis and sulfur cycle is being deeply studied at this moment. These studies will position AMEs as suitable extraterrestrial life search models.

References

Albarracín VH, Dib JR, Ordoñez OF, Farias M (2011) A harsh life to indigenous proteobacteria at the andean mountains : microbial diversity and resistance mechanisms towards extreme conditions. In: Sezenna ML (ed) Proteobacteria: phylogeny, metabolic diversity and ecological effects. Nova Science Publishers, Inc, pp 1–29

Albarracín VH, Pathak GP, Douki T, Cadet J, Borsarelli CD, Gärtner W, Farias ME (2012) Extremophilic Acinetobacter strains from high-altitude lakes in Argentinean Puna: remarkable UV-B resistance and efficient DNA damage repair. Orig Life Evol Biosph 42(2–3):201–221

Albarracín VH, Gärtner W, Farías ME (2013) UV resistance and photoreactivation of extremophiles from high-altitude Andean Lakes. J Photochem Photobiol B Biol:1–4

Albarracín VH, Kurth D, Ordoñez OF, Belfiore C, Luccini E, Salum GM, Piacentini RD, Farías ME (2015) High-up: a remote reservoir of microbial extremophiles in Central Andean wetlands. Front Microbiol 6:1404

Albarracín VH, Gärtner W, Farias ME (2016a) Forged under the sun: life and art of extremophiles from Andean Lakes. Photochem Photobiol 92:14–28

Albarracín VH, Kraiselburd I, Bamann C, Wood PG, Bamberg E, Farias ME, Gärtner W (2016b) Functional green-tuned proteorhodopsin from modern stromatolites. PLoS One 11:1–18

Allwood AC, Walter MR, Kamber BS, Marshall CP, Burch IW (2006) Stromatolite reef from the Early Archaean era of Australia. Nature 441:714–718

Anbar A, Davies PCW, Wolfe-Simon F (2009) Did nature also choose arsenic. Int J Astrobiol 8:69–74

Andres J, Bertin PN (2016) The microbial genomics of arsenic. FEMS Microbiol Rev 40:299–322

Belluscio A (2010) Hostile volcanic lake teems with life. Nature. https://doi.org/10.1038/news.2010.161

Benedetto JL (2010) El continente Gondwana a través del tiempo. Una introducción a la geología histórica. Academia Nacional de Ciencias, Córdoba, Argentina

Boutt DF, Hynek SA, Munk LA, Corenthal LG (2016) Rapid recharge of fresh water to the halite-hosted brine aquifer of Salar de Atacama, Chile. Hydrol Process 30:4720–4740

Bundschuh J, Gimene-Forcada E, Guerequiz E et al (2008) Fuentes geogénicas de arsénico y su liberación al medio ambiente. CYTED, Ciencia y Tecnología para el Desarrollo:33–47

Burne RV, Moore LS (1987) Microbialites: organosedimentary deposits of benthic microbial communities. PALAIOS 2:241

Cabral AR, Beaudoin G (2007) Volcanic red-bed copper mineralisation related to submarine basalt alteration, Mont Alexandre, Quebec Appalachians, Canada. Mineral Deposita 42:901–912

Clarke JDA (2006) Antiquity of aridity in the Chilean Atacama Desert. Geomorphology 73:101–114

Connon SA, Lester ED, Shafaat HS, Obenhuber DC, Ponce A (2007) Bacterial diversity in hyperarid Atacama Desert soils. J Geophys Res Biogeo 112(G4)

Contador CA, Veas-Castillo L, Tapia E, Antipán M, Miranda N, Ruiz-Tagle B, García-Araya J, Andrews BA, Marin M, Dorador C, Asenjo JA (2019) Atacama Database: a platform of the microbiome of the Atacama Desert. Antonie Van Leeuwenhoek:1–11

Corenthal LG, Boutt DF, Hynek SA, Munk LA (2016) Regional groundwater flow and accumulation of a massive evaporite deposit at the margin of the Chilean Altiplano. Geophys Res Lett 43:8017–8025

Davila AF, Gómez-Silva B, de Los RA, Ascaso C, Olivares H, McKay CP, Wierzchos J (2008) Facilitation of endolithic microbial survival in the hyperarid core of the Atacama Desert by mineral deliquescence. J Geophys Res Biogeosci 113(G1)

Demergasso CS, Guillermo CD, Lorena EG et al (2007) Microbial precipitation of arsenic sulfides in andean salt flats. Geomicrobiol J 24:111–123

Di Capua C, Bortolotti A, Farías ME, Cortez N (2011) UV-resistant *Acinetobacter* sp. isolates from Andean wetlands display high catalase activity. FEMS Microbiol Lett 317:181–189

Dong H, Rech JA, Jiang H, Sun H, Buck BJ (2007) Endolithic cyanobacteria in soil gypsum: occurrences in Atacama (Chile), Mojave (United States), and Al-Jafr Basin (Jordan) Deserts. J Geophys Res Biogeosci 112:1–11

Duffie JA, Beckman WA (2013) Solar engineering of thermal processes, 4th edn. Wiley, Hoboken

Dupraz C, Reid RP, Braissant O et al (2009) Processes of carbonate precipitation in modern microbial mats. Earth Sci Rev 96:141–162

Escudero LV, Casamayor EO, Chong G, Pedrós-Alió C, Demergasso C (2013) Distribution of microbial arsenic reduction, oxidation and extrusion genes along a wide range of environmental arsenic concentrations. PLoS One 8:e78890

Farías ME, Rascovan N, Toneatti DM et al (2013) The discovery of stromatolites developing at 3570 m above sea level in a high-altitude volcanic lake Socompa, Argentinean Andes. PLoS One 8:e53497

Farías ME, Contreras M, Rasuk MC et al (2014) Characterization of bacterial diversity associated with microbial mats, gypsum evaporites and carbonate microbialites in thalassic wetlands: Tebenquiche and La Brava, Salar de Atacama, Chile. Extremophiles 18:311–329

Farias ME, Rasuk MC, Gallagher KL, Contreras M, Kurth D, Fernandez AB, Poiré D, Novoa F, Visscher PT (2017) Prokaryotic diversity and biogeochemical characteristics of benthic microbial ecosystems at La Brava, a hypersaline lake at Salar de Atacama, Chile. PLoS One 12(11):e0186867

Fernández Zenoff V, Siñeriz F, Farías ME (2006) Diverse responses to UV-B radiation and repair mechanisms of bacteria isolated from high-altitude aquatic environments. Appl Environ Microbiol 72(12):7857–7863

Fernandez AB, Rasuk MC, Visscher PT, Contreras M, Novoa F, Poire DG, Patterson MM, Ventosa A, Farias ME (2016) Microbial diversity in sediment ecosystems (evaporites domes, microbial mats, and crusts) of hypersaline Laguna Tebenquiche, Salar de Atacama, Chile. Front Microbiol 7:1284

Fox GC, Stackebrandt E, Hespell RB, Gibson J, Maniloff J, Dyer TA, Wolfe RS, Balch WE, Tanner RS, Magrum LJ, Zablen LB (1980) The phylogeny of prokaryotes. Science 209(4455):457–463

Garreaud RD, Vuille M, Compagnucci R, Marengo J (2009) Present-day South American climate. Palaeogeogr Palaeoclimatol Palaeoecol 281:180–195

Glunk C, Dupraz C, Braissant O, Gallagher KL, Verrecchia EP, Visscher PT (2011) Microbially mediated carbonate precipitation in a hypersaline lake, Big Pond (Eleuthera, Bahamas). Sedimentology 58:720–736

Gomez FJ, Kah LC, Bartley JK, Astini RA (2014) Microbialites in a high-altitude Andean lake: multiple controls on carbonate precipitation and lamina accretion. PALAIOS 29:233–249

Hartley AJ, Chong G, Houston J, Mather AE (2005) 150 million years of climatic stability: evidence from the Atacama Desert, northern Chile. J Geol Soc Lond 163(3):421–424

Holland HD (1994) Early Proterozoic atmospheric change. In: Bengston S (ed) Early life on earth. Columbia University Press, New York, pp 237–244

Holland HD (1997) Evidence for life on Earth more than 3850 million years ago. Science 275:38–39

Kasting JF (2001) Earth history: the rise of atmospheric oxygen. Science 293(5531):819–820

Kempe S, Kazmierczak J (2002) Biogenesis and early life on Earth and Europa: favored by an alkaline ocean? Astrobiology 2(1):123–130

Kish MM, Viola RE (1999) Oxyanion specificity of aspartate-β-semialdehyde dehydrogenase. Inorg Chem 38:818–820

Knauth LP (1998) Salinity history of the Earth's early ocean. Nature 395:554–555

Kulp TR (2014) Arsenic and primordial life. Nat Geosci 7:785–786

Kulp TR, Hoeft SE, Asao M, Madigan MT, Hollibaugh JT, Fisher JC, Stolz JF, Culbertson CW, Miller LG, Oremland RS (2008) Arsenic(III) fuels anoxygenic photosynthesis in hot spring biofilms from Mono Lake, California. Science 321(5891):967–970

Kurth D, Belfiore C, Gorriti MF, Cortez N, Farias ME, Albarracín VH (2015) Genomic and proteomic evidences unravel the UV-resistome of the poly-extremophile Acinetobacter sp. Ver3. Front Microbiol 6:328

Kurth D, Amadio A, Ordoñez OF, Albarracín VH, Gärtner W, Farías ME (2017) Arsenic metabolism in high altitude modern stromatolites revealed by metagenomic analysis. Sci Rep 7:1024

Lahaye Y, Barnes S-J, Frick LR, Lambert DD (2001) Re-Os Isotopic study of komatiitic volcanism and magmatic sulfide formation in the Southern Abitibi Greenstone Belt, Ontario, Canada. Can Mineral 39(2):473–490

Larsen H (1980) Ecology of hypersaline environments. Dev Sedimentol 28:23–39

Lebrun E, Brugna M, Baymann F, Muller D, Lièvremont D, Lett MC, Nitschke W (2003) Arsenite oxidase, an ancient bioenergetic enzyme. Mol Biol Evol 20:686–693

Luccini E, Cede A, Piacentini R, Villanueva C, Canziani P (2006) Ultraviolet climatology over Argentina. J Geophys Res 111:D17312

Marshall KC (1984) Microbial adhesion and aggregation. In: Dahlem Konferenzen. Springer, Berlin

Mojzsis SJ, Arrhenius G, McKeegan KD, Harrison TM, Nutman AP, Friend CR (1996) Evidence for life on Earth before 3,800 million years ago. Nature 384:55–59

Nutman AP, Bennett VC, Friend CR, Van Kranendonk MJ, Chivas AR (2016) Rapid emergence of life shown by discovery of 3,700-million-year-old microbial structures. Nature 537:535–538

Nutman AP, Bennett VC, Friend CR, Van Kranendonk MJ, Rothacker L, Chivas AR (2019) Cross-examining Earth's oldest stromatolites: seeing through the effects of heterogeneous deformation, metamorphism and metasomatism affecting Isua (Greenland) ~3700 Ma sedimentary rocks. Precambrian Res 331:105347

Ordoñez OF, Rasuk MC, Soria MN, Contreras M, Farías ME (2018) Haloarchaea from the Andean Puna: biological role in the energy metabolism of arsenic. Microb Ecol 76(3):695–705

Oremland RS, Saltikov CW, Wolfe-Simon F, Stolz JF (2009) Arsenic in the evolution of Earth and extraterrestrial ecosystems. Geomicrobiol J 26(7):522–536

Parnell J, Lee P, Cockell CS, Osinski GR (2004) Microbial colonization in impact-generated hydrothermal sulphate deposits, Haughton impact structure, and implications for sulphates on Mars. Int J Astrobiol 3:247–256

Pla-García J, Menor-Salván C (1990) La composición química de la atmósfera primitiva del planeta Tierra. Real Sociedad Española de Química 113(7)

Rapf RJ, Vaida V (2016) Sunlight as an energetic driver in the synthesis of molecules necessary for life. Phys Chem Chem Phys 18:20067–20084

Rascovan N, Maldonado J, Vazquez MP, Eugenia Farías M (2016) Metagenomic study of red biofilms from Diamante Lake reveals ancient arsenic bioenergetics in haloarchaea. ISME J 10:299–309

Rasuk MC, Kurth D, Flores MR et al (2014) Microbial characterization of microbial ecosystems associated to evaporites domes of gypsum in salar de Llamara in Atacama Desert. Microb Ecol 68:483–494

Reid RP, Visscher PT, Decho AW, Stolz JF, Bebout BM, Dupraz C, Macintyre IG, Paerl HW, Pinckney JL, Prufert-Bebout L, Steppe TF (2000) The role of microbes in accretion, lamination and early lithification of modern marine stromatolites. Nature 406:989–992

Risacher F, Alonso H, Salazar C (2003) The origin of brines and salts in Chilean salars: a hydrochemical review. Earth Sci Rev 63:249–293

Rosing MT (1999) ^{13}C-depleted carbon microparticles in $\geq$3700-Ma sea-floor sedimentary rocks from west Greenland. Science 283:674–676

Rothschild L (1990) Earth analogs for Martian life. Microbes in evaporites, a new model system for life on Mars. Icarus 88:246–260

Sancar A (2003) Photolyase and cryptochrome blue-light photoreceptors. Chem Rev 103:2203–2237

Sancho-Tomás M, Somogyi A, Medjoubi K, Bergamaschi A, Visscher PT, Van Driessche AE, Gérard E, Farias ME, Contreras M, Philippot P (2018) Distribution, redox state and (bio) geochemical implications of arsenic in present day microbialites of Laguna Brava, Salar de Atacama. Chem Geol 490:13–21

Sanz-Montero ME, Rodríguez-Aranda JP (2008) Participación microbiana en la formación de magnesita dentro de un ambiente lacustre evaporítico: Mioceno de la Cuenca de Madrid. Macla 9:231–232

Schulze-Makuch D, Wagner D, Kounaves SP, Mangelsdorf K, Devine KG, de Vera JP, Schmitt-Kopplin P, Grossart HP, Parro V, Kaupenjohann M, Galy A (2018) Transitory microbial habitat in the hyperarid Atacama Desert. Proc Natl Acad Sci USA 115:2670–2675

Shih PM (2015) Photosynthesis and early Earth. Curr Biol 25:R855–R859

Smedley PL, Kinniburgh DG (2002) A review of the source, behaviour and distribution of arsenic in natural waters. Appl Geochem 17:517–568

Summons RE, Jahnke LL, Hope JM, Logan GA (1999) 2-Methylhopanoids as biomarkers for cyanobacterial oxygenic photosynthesis. Nature 400:554–557

Tawfik DS, Viola RE (2011) Arsenate replacing phosphate: alternative life chemistries and ion promiscuity. Biochemistry 50:1128–1134

Thompson JB, Ferris FG (1990) Cyanobacterial precipitation of gypsum, calcite, and magnesite from natural alkaline lake water. Geology 18:995–998

Van Kranendonk MJ, Philippot P, Lepot K, Bodorkos S, Pirajno F (2008) Geological setting of Earth's oldest fossils in the ca. 3.5 Ga Dresser Formation, Pilbara Craton, Western Australia. Precambrian Res 167:93–124

Vasconcelos C, McKenzie JA (1997) Microbial mediation of modern dolomite precipitation and diagenesis under anoxic conditions (Lagoa Vermelha, Rio de Janeiro, Brazil). J Sediment Res 67:378–390

Vuille M, Bradley RS, Keimig F (2000) Interannual climate variability in the Central Andes and its relation to tropical Pacific and Atlantic forcing. J Geophys Res Atmos 105:12447–12460

Vuille M, Keimig F, Vuille M, Keimig F (2004) Interannual variability of summertime convective cloudiness and precipitation in the Central Andes derived from ISCCP-B3 data. J Clim 17:3334–3348

Walter MR, Buick R, Dunlop JSR (1980) Stromatolites 3,400-3,500 Myr old from the North Pole area, Western Australia. Nature 284:443–445

Watanabe Y, Martini JE, Ohmoto H (2000) Geochemical evidence for terrestrial ecosystems 2.6 billion years ago. Nature 408:574–578

Weber S (2005) Light-driven enzymatic catalysis of DNA repair: a review of recent biophysical studies on photolyase. Biochim Biophys Acta Bioenerg 1707(1):1–23

Weiss MC, Sousa FL, Mrnjavac N, Neukirchen S, Roettger M, Nelson-Sathi S, Martin WF (2016) The physiology and habitat of the last universal common ancestor. Nat Microbiol 1(9):16116

Witt-Eickschen G, Palme H, O'Neill HSC, Allen CM (2009) The geochemistry of the volatile trace elements As, Cd, Ga, In and Sn in the Earth's mantle: new evidence from in situ analyses of mantle xenoliths. Geochim Cosmochim Acta 73:1755–1778

Wolfe-Simon F, Blum JS, Kulp TR, Gordon GW, Hoeft SE, Pett-Ridge J, Stolz JF, Webb SM, Weber PK, Davies PC, Anbar AD (2011) A bacterium that can grow by using arsenic instead of phosphorus. Science 332(6034):1163–1166

Chapter 9
Stromatolites in Crater-Lake Alchichica and Bacalar Lagoon

Luisa I. Falcón, Patricia M. Valdespino-Castillo, Rocio J. Alcántara-Hernández, Elizabeth S. Gómez-Acata, Alfredo Yanez-Montalvo, and Bernardo Águila

Abstract Extant stromatolites have been considered ecological similes to their ancient counterparts. We now know that these microbial assemblages are composed of a great diversity of microbes, which couple and intertwine their metabolic capabilities to create self-sustained microbial ecosystems. Presently, stromatolites thrive in a vast diversity of aquatic environments including freshwater, hypersaline, coastal lagoons, alkaline lakes, oligotrophic pools, abandoned pits, few marine systems, and brackish waters. In this chapter, we will summarize the research that has been done in two stromatolite-harboring sites in Mexico: the alkaline crater-lake Alchichica and the oligotrophic karstic coastal lagoon of Bacalar. Alchichica is located in the Transvolcanic belt in Central Mexico. It is a maar-alkaline crater lake (salinity 8.5 gl^{-1}, pH 9.5) with water chemistry determined by high contents of carbonates, sodium, a high Mg/Ca ratio, and particularly low Ca^{2+} concentrations (~0.3 mM). Two main stromatolite-types, as defined by mineralogy, texture, and microbial composition, develop along its periphery from surface to over 30 m in depth. Alchichica is a modern environment that resembles Precambrian oceanic conditions. Stromatolites from Alchichica have been dated radiometrically in ~1.1–2.8 ka BP. Bacalar is a coastal lagoon located in the Yucatan Peninsula, which is a carbonate platform that emerged above sea level during the Oligocene. The Bacalar lagoon has high carbonate concentrations as a result of the influx of ground-

L. I. Falcón (✉) · E. S. Gómez-Acata · A. Yanez-Montalvo · B. Águila
Instituto de Ecología, Universidad Nacional Autónoma de México, UNAM, Parque Científico y Tecnológico de Yucatán, Mérida, Yucatán, Mexico
e-mail: falcon@ecologia.unam.mx; esga@iecologia.unam.mx; ayanez@ecosur.edu.mx

P. M. Valdespino-Castillo
MBIB, Lawrence Berkeley National Laboratory, Berkeley, USA
e-mail: pmvaldespino@lbl.gov

R. J. Alcántara-Hernández
Instituto de Geología, Universidad Nacional Autónoma de México, UNAM, Ciudad Universitaria, Mexico City, Mexico
e-mail: ralcantarah@geologia.unam.mx

© Springer Nature Switzerland AG 2020
V. Souza et al. (eds.), *Astrobiology and Cuatro Ciénegas Basin as an Analog of Early Earth*, Cuatro Ciénegas Basin: An Endangered Hyperdiverse Oasis, https://doi.org/10.1007/978-3-030-46087-7_9

183

water (salinity 1.2 gl⁻¹, pH 7.6–8.3). It holds the largest freshwater stromatolite structures known, which have been radiometrically dated in ~6.8–9.2 ka BP. Their mineralogy, shape, and texture are similar along the lagoon's coast, but microbial composition changes, possibly due to anthropogenic impact.

9.1 The Ecological Setting

9.1.1 Crater-Lake Alchichica

Alchichica is an oligotrophic, monomictic, saline, maar, soda, crater-lake located in the Trans-Mexican Volcanic Belt in central Mexico (19° 24' N, 97° 24' W, 2300 m.a.s.l) (Fig. 9.1). It has a maximum depth of 60 m, contains alkaline water (pH 8.7–9.2), and a high conductivity (13 dS m⁻¹). The dominant anions are chlorides and bicarbonates; whereas the dominant cations are sodium and magnesium (Alcocer and Lugo 2003; Armienta et al. 2008; Mancilla Villa et al. 2014). Kaźmierczak et al. (2011) described two distinct morphologic generations of stromatolites: (1) Columnar dome-like structures, rich in aragonite, forming near the shore line, dated back to 1100 ybp and (2) Spongy cauliflower-like thrombolytic structures, rich in hydromagnesite and aragonite, dominate the lake's periphery (Valdespino-Castillo et al. 2018) (Fig. 9.2). Developing inside the lake, these structures were dated back to 2800 ybp (Kaźmierczak et al. 2011). It is hypothesized that these distinct morphologies and mineral compositions were the result of different periods of drought and flood by groundwater influx to the lake.

9.1.2 Bacalar Lagoon

Bacalar is the largest freshwater karstic lake in the Yucatan Peninsula (Gamboa-Pérez and Schmitter-Soto 1999) (Fig. 9.1). The lagoon is 40 km long, 1–2 km wide, and 20 m at its deepest point (Gischler et al. 2008). It is located in the evaporitic region of the Yucatan Peninsula, in an area defined as a "Coastal Transversal Corridor," due to its high hydrological connectivity (Perry et al. 2002; Hernández-Arana et al. 2015). Bacalar lagoon has a high groundwater inflow and connections with other water bodies – brackish lagoons, estuarine coastal lagoons, and reef lagoons – especially during the rainy season (Perry et al. 2009; Hernández-Arana et al. 2015). The physicochemical characteristics of the water are oligotrophic, alkaline (7.5–8.3) with a temperature range of 28–31 °C (Gamboa-Pérez and Schmitter-Soto 1999; Gischler et al. 2008; Beltrán et al. 2012; Centeno et al. 2012). The hydrogeochemistry has been defined as $Ca^{2+} > Mg^{2+} > Na^+ > Sr^{2+} > K^+$. The abundance of anions is $SO_4^{2-} >> HCO_3^- > Cl^- > NO_3^-$ (Gischler et al. 2008; Castro-Contreras et al. 2014; Velázquez et al. 2018). Sulfates show saturation throughout

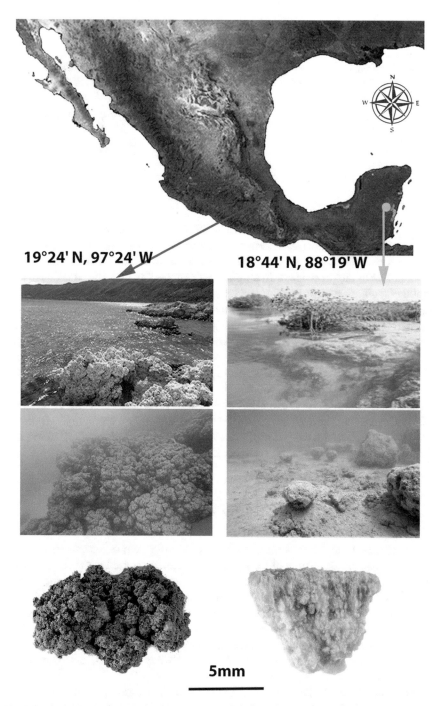

Fig. 9.1 Geographic location of stromatolites in Mexico. Crater-lake Alchichica (left) within the central Mexican Transvolcanic Belt, and Bacalar lagoon (right) in southern Yucatan peninsula

Fig. 9.2 Mineral composition of crater-lake Alchichica stromatolite types (AS, spongy; AC, columnar) and Bacalar lagoon stromatolites (BAC). Modified from Valdespino-Castillo et al. 2018

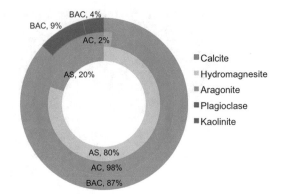

the lagoon (Perry et al. 2009; Johnson et al. 2018). Calcium is higher in the northern zone, with values similar to seawater (Gischler et al. 2008). A saturation of bicarbonate ions, with values greater than seawater, occurs in the southern part of the lagoon, due to the presence of four sinkhole "cenotes." In addition, it is considered that the southern part of Bacalar lagoon has the greatest extent of freshwater stromatolites (~ 10 km) in the world (Gischler et al. 2008, 2011). Stromatolites in Bacalar lagoon are distributed in reef patches near the shores throughout the lagoon. These stromatolites have been radiometrically dated between 6790 and 9190 ybp (Gischler et al. 2008). Their morphology varies with the depth of the shore (Gischler et al. 2011). They have been described as forming crusts, heads, and ledges. The internal texture of Bacalar stromatolites suggests that most of them are thrombolites, but there are also stromatolites and oncolites present (Gischler et al. 2008, 2011; Castro-Contreras et al. 2014). Bacalar stromatolites are composed almost entirely of calcite, with a mixture of plagioclase, kaolinite, and siderite (Valdespino-Castillo et al. 2018; Yanez-Montalvo et al. 2020) (Fig. 9.2).

9.2 Molecular Approaches to the Study of Stromatolites

Stromatolites have been studied with mineral, biochemical, microscopic, and genetic protocols. The study of their microbial diversity started with morphological descriptions using different microscopic techniques (i.e., light microscopy, scanning electron microscopy (SEM), transmission electron microscopy (TEM), and laser confocal microscopy). Molecular biology and Next Generation Sequencing techniques (NGS) have allowed the characterization of microbial communities through gene amplification from environmental DNA (Centeno et al. 2012; Coman et al. 2015; Lindsay et al. 2017; Valdespino-Castillo et al. 2018; Gómez-Acata et al. 2019).

Studying stromatolites with molecular techniques started in this century by Steppe et al. (2001), who were the first to extract DNA from microbial communities in stromatolites. Several different techniques have been used where the principal objective is to obtain DNA of good quality and in great quantity for downstream analysis such as metagenomic studies (e.g., Gómez-Acata et al. 2019). All of these methods have three principal steps: a breakage of both cell wall and cell membranes with different approaches including mechanical disruption, physical, chemical, and enzymatic lysis; hydrophobic cell debris separation (proteins and lipids are eliminated with organic solvents such as phenol:chloroform and DNA precipitation using alcohols. Moreover, depending on the type of sample, some methods include other steps to eliminate any inhibitors produced by microorganisms (i.e., exopolymeric substances (EPS) and carbonated minerals) at the beginning of the DNA extraction method (pre-treatment) or after DNA precipitation (purification) (Gómez-Acata et al. 2019). Commercial kits (used for environmental samples such as soil, plants, sediments, and biofilm) are a good choice since these are efficient and quick DNA extraction methods (López-García et al. 2005; Breitbart et al. 2009; Couradeau et al. 2011; Saghaï et al. 2015; Lindsay et al. 2017). There is no best method for DNA extraction of stromatolites; all are good and have success depending on the mineral composition and the amount of EPS in the sample. Methods that could be inadequate for one sample could be the best for another, and the method selection depends on the stromatolite degree of lithification.

Microbialites have also been studied through the analysis of RNA expression. For this purpose, it is very important to isolate high-quality RNA, so elimination of all nucleases in the environment is crucial for this purpose since nucleases degrade RNA. Phenol is a very common reagent used for elimination of nucleases in RNA extraction methods as well as β-mercaptoethanol and guanidine thiocyanate (Santos et al. 2010; Mobberley et al. 2015; Louyakis et al. 2018). Furthermore, commercial kits have been used with excellent results (Edgcomb et al. 2014; Alcántara-Hernández et al. 2017; Valdespino-Castillo et al. 2014, 2017). When analyzing gene expression patterns, sampling is fundamental, since enzymatic activity changes with the time of the day; for this reason, the sample must be frozen immediately with liquid nitrogen at the study site or when the study site is difficult to access, some preservative solution could be used (i.e., RNA later) (Kruse et al. 2017). Finally, the integrity of RNA must be determined through the RIN (RNA integrity number) value and the sample must be free of any inhibitor. There are a few studies of phages (viruses) in stromatolites (Desnues et al. 2008; White III et al. 2018), by direct methods (isolation of viruses from samples purified with a cesium chloride density gradient and formamide/CTAB extraction).

Alchichica crater-lake and Bacalar lagoon stromatolites have been studied to understand their microbial community composition with genomic approaches. For this purpose, DNA extractions have included phenol:chloroform with enzymatic lysis, mechanical disruption, and chemical lysis (Centeno et al. 2012; Valdespino-Castillo et al. 2014, 2017, 2018; Yanez-Montalvo et al. 2020) or using DNeasy PowerSoil® (Couradeau et al. 2011) and Power Biofilm TM DNA isolation kits (Qiagen, Carlsbad, CA, USA) (Ragón et al. 2014). In some studies, initial carbonate dissolution was needed to obtain high-quality DNA (Table 9.1), suitable for downstream applications. Moreover, Alchichica crater-lake stromatolites have been

Table 9.1 DNA extraction techniques for microbial community studies in stromatolites from the Bacalar lagoon and the Alchichica crater-lake

Study site	Lysis treatment	Technique/commercial kit	DNA purification method[a]	Reference
Alchichica crater-lake	Carbonates dissolved with HCl 33% and neutralized with PBS pH 7, 0.5 M EDTA pH 9	(1) QuickPick™gDNA Kit (Bio-Nobile, Parainen, Finland) previous incubation for 3 h at 56 °C with Proteinase K and ViscozymeH	None	Couradeau et al. (2011)
		(2) MoBioPowerSoil DNA kit (MoBio, Carlsbad, CA, USA) previous incubation with ViscozymeH (Sigma-Aldrich, Buchs, Switzerland)		
Bacalar lagoon and Alchichica crater-lake	None	Chemistry lysis with CTAB and SDS, enzymatic lysis with lysozyme, freeze-thaw cycles, and mortar liquid nitrogen, extraction with phenol:chloroform:isoamyl alcohol, precipitation with isopropanol	DNeasy Blood & Tissue kit (Qiagen, Alameda, CA)	Centeno et al. (2012)
Alchichica crater-lake	None	Power Biofilm ™ DNA Isolation Kit (MoBio, Carlsbad, CA, USA)	None	Ragon et al. (2014)
Alchichica crater-lake	None	DNA was extracted as reported by Centeno et al. (2012)	None	Valdespino-Castillo et al. (2014, 2017, 2018) and Alcántara-Hernández et al. (2017)
Alchichica crater-lake	Carbonates dissolved with 100 µL HCl 33% then neutralized with 1 mL of 1:1 PBS pH 7 and 0.5 M EDTA pH 9	Ground with mortar and pestle and Power Biofilm ™ DNA Isolation Kit (MoBio, Carlsbad, CA, USA)	Power Clean™ Pro DNA Clean-up Kit (Mo Bio, Carlsbad, CA, USA), to remove residual EPS	Saghaï et al. (2015)

[a]Methods that report an extra purification step after DNA extraction

studied with an RNA, cDNA approach (Valdespino-Castillo et al. 2017; Alcántara-Hernández et al. 2017). The extraction was done with an RNA PowerSoil® Total RNA Isolation kit (Mo Bio Laboratories, Carlsbad, CA). Then RNA was purified with an RNeasy Mini kit (Qiagen, Venlo, the Netherlands) and DNA digestion with DNAseI (Qiagen) was necessary.

9.3 Microbial Diversity

Stromatolite formation is due to mineral accretion, precipitation, or binding influenced by microbial activity (Burne and Moore 1987). The main biological components of stromatolites are bacteria, which represent over 90% of genetic diversity (Centeno et al. 2012). Research suggests that stromatolites are formed by a microbial succession, where microbial mats actively trap ions and detritus with mucous EPS. This EPS forms nucleation centers, and when they become saturated, an extracellular precipitation aided by microbial degradation occurs (Dupraz et al. 2009). Although no concise bacterial metabolisms have been reported for microbialite genesis, some metabolisms are most likely to influence precipitation, cementation, and lithification of minerals (Riding 2011). For instance, the role of photosynthesis, either oxygenic and/or anoxygenic, along with sulfur metabolism is key for mineral precipitation (Arp et al. 2012). Photosynthesis elevates pH locally and increases the saturation index of minerals in a microenvironment, thus forming a genesis or growth by lamination (Gérard et al. 2013). Other metabolisms may be involved in carbonate precipitation, including ureolysis, nitrification, and methane metabolism (Zhu and Dittrich 2016). A commonality among stromatolites is the presence of cyanobacteria, sulfate-reducing bacteria, and Proteobacteria. This bacterial composition seems to be key for the construction of stromatolites (Chagas et al. 2016).

9.3.1 Microbial Diversity of Alchichica Crater-Lake Stromatolites

Studies in crater-lake Alchichica have described the microbial diversity of stromatolites in both stromatolite-types, around the lake's periphery (Centeno et al. 2012) and along depth gradients (Kaźmierczak et al. 2011; Couradeau et al. 2011, Águila 2018). Stromatolite colorations vary with depth; for instance, shallow microbialites have a yellow-greenish coloration, while deeper microbialites exhibit brown-reddish colorations, due to different photopigments (Kaźmierczak et al. 2011; Couradeau et al. 2011; Águila 2018). Overall, both stromatolite types (spongy and columnar) have a strong component of Cyanobacteria (Nostocales, Oscillatoriales, and Synechococcales), Proteobacteria (mainly Alphaproteobacteria, followed by Deltaproteobacteria), Bacteroidetes (mainly Cythophagia) and Actinobacteria (Fig. 9.3) (Valdespino-Castillo et al. 2018).

9.3.1.1 Cyanobacteria of Alchichica Crater-Lake Stromatolites

Alchichica stromatolites contain a high diversity and abundance of cyanobacteria, including several unique species (Tavera and Komárek 1996). A recent study of cyanobacterial composition along a depth gradient (from 3 m to 30 m) found that

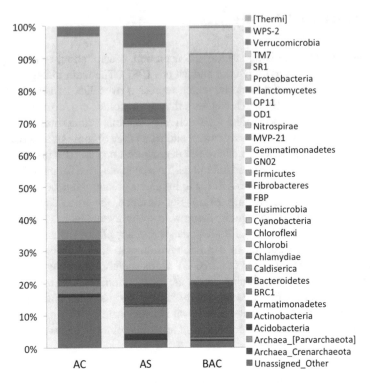

Fig. 9.3 Microbial composition based on 16SrDNA gene diversity associated with Alchichica crater-lake (AC, columnar; AS, spongy) and Bacalar lagoon (BAC) stromatolites

Pleurocapsales, and particularly the Xenococcaceae family, are the main cyano-bacteria present in deep stromatolites below 12 m (Águila 2018). These results agree with previous studies (Couradeau et al. 2011; Kaźmierczak et al. 2011), where a taxonomy/depth relationship was observed. For instance, Nostocales and filamentous Synechococcales were found at shallow depths, while Pleurocapsales and Chroococcales were associated with deeper stromatolites; no correlation was found with cocci Synechococales. Interestingly, stromatolites from 30 m only contained *Chroococcidium gelatinosum* and *Xenococcus candelaria*, suggesting an important role in their formation (Águila 2018). The main cyanobacterial gen-era described in Alchichica crater-lake stromatolites included: *Rivularia, Calothrix, Trichormus, Nodosilinea, Phormidesmis, Oculatella, Cyanobium, Heteroleiblena, Chroococcus, Entophysalis, Chroococcopsis, Xenococcus, Chroococcidium, Stanieria sp., Pseudanabaena, Synechococcus,* and *Gloeomargarita* along with a high diversity of unclassified cocci Synechococales (Águila 2018) (Fig. 9.4).

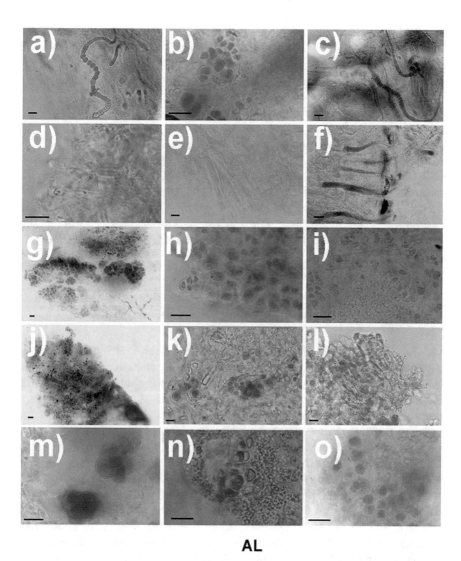

AL

Fig. 9.4 Representative images of the cyanobacterial diversity of crater-lake Alchichica. Several classes of cyanobacteria are observed including Nostocales (**a**, **c** and **f**), Synechococcales (**d** and **e**), Chroococcales (**d**, **h**, **i** and **o**), and Pleurocapsales (**b**, **g**, **j**, **k**, **l**, **m** and **n**). Scale bar is 5 μm

9.3.2 Microbial Diversity of Bacalar Lagoon

Bacalar stromatolites have shown a high content of C_{org} associated with a large cyanobacterial abundance, representing over 70% of their genetic diversity, followed by Bacteroidetes and Proteobacteria (Valdespino-Castillo et al. 2018; Yanez-Montalvo 2020) (Fig. 9.3). In addition to the above groups, Johnson et al. (2018)

described that reducing sulfate bacteria contribute significantly to the precipitation of carbonates. Proteobacteria are represented primarily by classes Alpha, Beta, and Deltaproteobacteria (Fig. 9.3).

9.3.2.1 Cyanobacteria of Bacalar Lagoon Stromatolites

Cyanobacteria in stromatolites from Bacalar lagoon are common and diverse at the northern portion of the lagoon while the southern portion exhibits a low diversity of cyanobacteria (Yanez-Montalvo et al. 2020). Main cyanobacteria dominating the stromatolites comprise Nostocales along with filamentous Synechococcales. Main genera include *Arthrospira, Nostoc, Rivularia, Calothrix, Mastigocladopsis, Fischerella, Microcoleus, Phormidium, Aphanocapsa, Microcystis, Synechococcus, Cyanobium, Nodosilinea, Chroococcidopsis,* and *Staneria* (Fig. 9.5).

9.4   Functional Ecology and Biogeochemical Cycling

As mentioned above, stromatolites harbor a rich and diverse microbial community that also results in a diverse metabolic profile, including phototrophs, chemotrophs, denitrifiers, sulfur oxidizers, sulfate-reducers, and P-cycling bacteria, among others. These microbes are interrelated in highly complex metabolic networks within physicochemical gradients at the microscale level in the stromatolite microstructure (Wong et al. 2015). Metagenomic studies have confirmed this plethora of biogeochemical pathways (Breitbart et al. 2009; Casaburi et al. 2016; Khodadad and Foster 2012; Kurth et al. 2017; Ruvindy et al. 2016; Warden et al. 2016), where the main processes associated with elemental cycling (e.g., C, N, S, and P) are present, suggesting thus functional redundancy. In addition, ecophysiological determinations have also shown relevant metabolic features of microbialites (Beltrán et al. 2012; Khodadad and Foster 2012). However, it seems that the environmental and geochemical conditions of the surrounding media determine the main metabolic pathways used by these benthic communities (Breitbart et al. 2009; Kurth et al. 2017; Ruvindy et al. 2016). Here, we will review the contributions regarding the N-cycle and P-cycling processes in stromatolites from crater-lake Alchichica and Bacalar lagoon, as stromatolites are known to often occur in oligotrophic environments where microbial activity is largely restricted by nutrient availability, such as N or P (Pepe-Ranney et al. 2012).

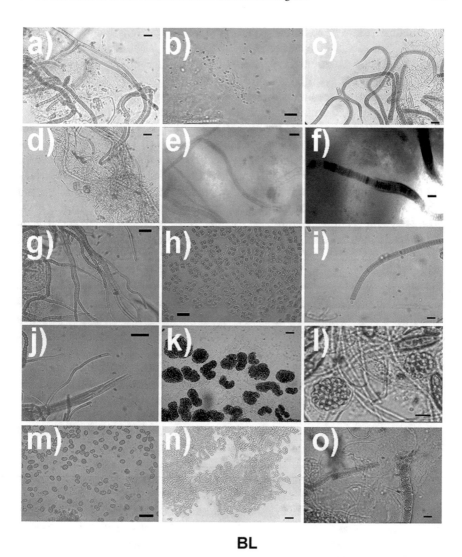

BL

Fig. 9.5 Representative images of the cyanobacterial diversity of Bacalar lagoon stromatolites. In Bacalar, observed classes of cyanobacteria include Nostocales (**c, f, g** and **o**) and Synechococcales (**a, b, d, e, h, i, j, k, l, m** and **n**). Scale bar is 5 μm

9.4.1   N-Cycle

Nitrogen is an essential element for biomolecules and its obtainment is essential for cellular development and growth. Consequently, this macroelement often limits biomass activity, where fixed-N might have constrained primary production on geological timescales (Falkowski 1997). Nitrogen is used in both assimilatory and dissimilatory pathways including organic N recycling reactions and redox transformations of dissolved inorganic nitrogen species (NH_4^+, NO_3^-, and NO_2^-); all of them interrelated in assimilatory, energy-obtaining, and energy-dissipating processes (Canfield et al. 2010).

Alchichica crater-lake and Bacalar lagoon are reported as meso-oligotrophic systems regarding the concentration of dissolved inorganic nitrogen (DIN) [Alchichica ~1.5 µM and Bacalar ~0.5 µM], where NO_3^- is the most representative dissolved N-form in the oxic system. Their salinity varies in: (i) the origin of the ions [Alchichica, athalassohaline inland soda lake and Bacalar, karstic inland-coastal lagoon]; (ii) their type [Alchichica Na > Mg > K > Ca and Bacalar [Ca > Mg > Na]; and (iii) their concentration [Alchichica electrical conductivity 13 dS/m, Bacalar 2.2 dS/m]. Despite these differences, the stromatolites from both aquatic systems can fix N_2, mainly during daytime (Falcón et al. 2002; Beltrán et al. 2012) and count with diazotrophic bacteria, such as Cyanobacteria and Alphaproteobacteria, as revealed by 16S rRNA and *nifH* gene surveys (Beltrán et al. 2012; Centeno et al. 2012). N_2 fixation has been found as a relevant feature in lithifying and non-lithifying microbial mats (Beltrán et al. 2012), though stromatolites with excessive NO_3^- availability can diminish N_2-fixation rates due to the preference of nitrate assimilation over N_2 fixation (Breitbart et al. 2009). Metagenomic data suggest that the assimilatory pathways are the most relevant in the stromatolite community, including amino acid metabolism, allantoin degradation, nitrate assimilation, and N_2 fixation (Breitbart et al. 2009; Khodadad and Foster 2012; Ruvindy et al. 2016). A study considering different types of stromatolites, where Bacalar lagoon and Alchichica crater-lake samples were included, showed that the overall microbial distribution patterns show a significant correlation with the N concentration in the living tissue (Valdespino-Castillo et al. 2018), demonstrating thus, the relevance of N in these microbial communities.

Denitrification is one of the most studied anaerobic respiration processes in stromatolites and microbial mats (Joye and Paerl 1994; Alcántara-Hernández et al. 2017), where nitrate (NO_3^-) is subsequently reduced to N_2 (NO_3^-, NO_2^-, NO, N_2O, N_2). This process is often used by microorganisms living in aerobic/anaerobic interphases where O_2 concentration fluctuates and an alternative electron acceptor is needed. As a consequence, stromatolites harbor a rich denitrifying community, especially when nitrate is available, as mentioned above. We surveyed the potential nitrite respiring bacterial communities in stromatolites from Alchichica crater-lake and Bacalar lagoon employing the genes encoding for nitrite reductases (*nirK* and *nirS*) (Fig. 9.6). Our results showed that both microbial communities harbor diverse phylotypes (at 5% nucleotide difference), with a main proportion of shared OTUs (highlighted in green). Most of the *nirK* and *nirS* sequences were affiliated to Alphaproteobacteria; however, gene *nirS* OTUs were related to *Thioalkalivibrio nitratireducens*, a lithotrophic sulfur-oxidizing bacteria. These results suggest that main phylotypes for nitrate respiration are present in both stromatolites, with differences in the less abundant sequences. Denitrification has been reported in other metagenomic studies (Breitbart et al. 2009), and found as a shared feature among stromatolites.

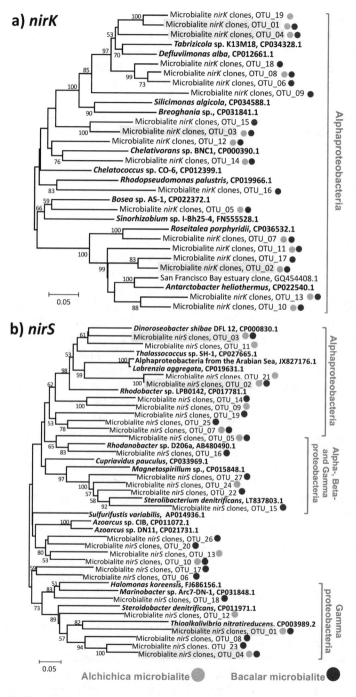

Fig. 9.6 Neighbor joining tree for genes involved in nitrite reduction in Alchichica lake and Bacalar lagoon microbialites. (**a**) *nirK* gene (469 nucleotides) and (**b**) *nirS* gene (402 nucleotides). Highlighted OTUs represent those with most clones (>5%) obtained in this study. Midpointed trees with bootstrap values ≥50% (1000 replicates). Divergence is represented by each scale bar

The chemolithotrophic processes such as nitrification and anaerobic ammonia oxidation (anammox) have showed differences among freshwater and marine stromatolites. In the Alchichica crater-lake, just a few phylotypes associated with aerobic and anaerobic ammonia oxidation were found, mainly related to *Nitrosomonas* spp. (capable of growth in slightly saline conditions). This low diversity could be related to bioenergetic constraints in this saline (alkaline) environment (Oren 2011), where nitrification and anammox are limited. The stromatolites of Bacalar lagoon have shown the presence of Thaumarchaeota, which likely drive ammonia oxidation in this system (Centeno et al. 2012).

9.4.2 P-Cycling

Phosphorus is a structural element of cell membranes and it is also interrelated to many other energetic processes in cellular function as well as an essential element for DNA and RNA construction. Microbes play a fundamental role in acquisition and remineralization of phosphorus within the ecosystems. In environmental sciences, we are still in the process of understanding the identity of the microorganisms harboring the genetic potential for the transformation of the different forms of phosphorus. In the aquatic environment, dissolved organic phosphorus (DOP) is usually the most abundant fraction (Kolowith et al. 2001) still largely uncharacterized (Dyhrman et al. 2007). Using stromatolites and bacterioplankton as study models, we have analyzed their potential to use this phosphorus compartment. Additionally, we compared stromatolites' potential to that of bacterioplankton.

To utilize this fraction of the total P in aquatic environments, microbes harbor diverse strategies, the better known are associated with pH-dependent (acid or alkaline) phosphatases to release orthophosphate or polyphosphate chains by the degradation of organic matter phosphate esters (Ruttenberg and Dyhrman 2005). Phytate or phytic acid is another molecule that may be abundant in DOP, it is associated overall with plant debris runoff to the aquatic environments (Suzumura and Kamatani 1995). Some microbes can release orthophosphate from phytate using specific enzymes, phytases. Following this logic, the microbial markers we used to study microbial DOP utilization in stromatolites and bacterioplankton of the maar lake Alchichica were alkaline phosphatases *phoA* and *phoD*, as well as beta propeller phytases (3-phytase, *BPP*). Finally, we studied the transcription of these genetic markers in seasonal and diel dynamics (Valdespino-Castillo et al. 2014, 2015, 2017).

Our results showed a large potential for DOP utilization using these specific genetic markers, taxonomically related to calcium-dependent alkaline phosphatases of Alpha-, Beta-, and Gamma-proteobacteria. A comparison with available datasets (GenBank-2014) showed that the diversity of alkaline phosphatase *phoX* of Alchichica was similar to the diversity of this particular marker in systems exhibiting contrasting environmental characteristics (e.g., trophic state, phosphorus condition, concentration of Zn, Mg, and Ca, metal cofactors of these type of enzymes) (Valdespino-Castillo et al. 2014). The diversity of BPP was taxonomically related to

Alpha- and Beta-proteobacteria, as well as to Bacteroidetes and Flavobacteriales. Beta propeller phytases found in this alkaline lake were similar to other phytases found in forest soils, glaciers, and lake sediments as well as from the gut of lake fish (Huang et al. 2009). These comparisons intend to contribute to understand the diversity of these bacterial metalloenzymes in natural waters; other markers should be compared among different habitats as the databases for microbial functional genes expand. A functional parallelism was found between stromatolites and bacterioplankton regarding DOP utilization where stromatolites concentrated most of the genetic diversity associated with the markers studied (*phoX, phoD,* and *BPP*) (Valdespino-Castillo et al. 2014). Alkaline phosphatase *phoX* was the marker among DOP utilization enzymes that showed the highest ratio of unique sequences/total sequences (Valdespino-Castillo et al. 2014).

Gene dynamics analyses showed a more stable pool of enzymes for DOP utilization in stromatolites compared to bacterioplankton, in which DOP utilization potential was relevantly impacted by the hydrologic cycles of the lake (summer stratification-winter circulation). The highest DOP utilization diversity of stromatolites together with this result suggests their resilience. The microbial communities in the water column exhibited a faster response to environmental change (different phylotypes were found in the intra-annual exploration (Valdespino-Castillo et al. 2014, 2017)). Together these results also suggest divergent strategies to face environmental change between stromatolites and bacterioplankton. An environment able to harbor high structural (and therefore functional) diversity might be a strategy that allows stromatolites to persist after millions of years. In terms of phosphorus utilization, stromatolite microstructure may offer a gradient of pH microenvironments (i.e., gradients that may be stretched by redox activity, such as photosynthesis and respiration), but the clarification of these gradients has been difficult to assess due to their rocky nature.

More studies are still needed to comprehend the intricacies of stromatolites in Alchichica crater-lake and Bacalar lagoon. These systems are unique and fragile. Both of them are experimenting a detriment in their water quality, associated with lack of wastewater treatment, and sustainable development. So far, we do not fully understand the rate of damage to these ancient structures, hence increasing our concern toward protecting these singular ecosystems. We now know how diverse and relevant stromatolites are functioning as micro-ecosystems where all metabolisms needed for their growth and maintenance exist, thanks to their great bacterial diversity.

References

Águila B (2018) Caracterización de cianobacterias en microbialitas del lago cráter Alchichica en un gradiente de profundidad. Tesis de MC, Posgrado en Ciencias Biológicas, Universidad Nacional Autónoma de México, UNAM, Ciudad de México, Mexico

Alcántara-Hernández RJ, Valdespino-Castillo PM, Centeno CM, Alcocer J, Merino-Ibarra M, Falcón LI (2017) Genetic diversity associated with N-cycle pathways in microbialites from Lake Alchichica, Mexico. Aquat Microb Ecol 78:121–133

Alcocer J, Lugo A (2003) Effects of El Niño on the dynamics of Lake Alchichica, central Mexico. Geofis Int 42(3):523–528

Armienta MA, Vilaclara G, De la Cruz-Reyna S, Ramos S, Ceniceros N, Cruz O, Aguayo A, Arcega-Cabrera F (2008) Water chemistry of lakes related to active and inactive Mexican volcanoes. J Volcanol Geotherm Res 178(2):249–258

Arp G, Helms G, Karlinska K, Schumann G, Reimer A, Reitner J, Trichet J (2012) Photosynthesis versus exopolymer degradation in the formation of microbialites on the atoll of Kiritimati, Republic of Kiribati. Cent Pac Geomicrobiol J 29(1):29–65

Beltrán Y, Centeno CM, García-Oliva F, Legendre P, Falcón LI (2012) N_2 fixation rates and associated diversity (*nifH*) of microbialite and mat-forming consortia from different aquatic environments in Mexico. Aquat Microb Ecol 67(1):15–24

Breitbart M, Hoare A, Nitti A, Siefert J, Haynes M, Dinsdale E, Edwards R, Souza V, Rohwer F, Hollander D (2009) Metagenomic and stable isotopic analyses of modern freshwater microbialites in Cuatro Ciénegas, Mexico. Environ Microbiol 11:16–34

Burne RV, Moore LS (1987) Microbialites: organosedimentary deposits of benthic microbial communities. PALAIOS:241–254

Canfield DE, Glazer AN, Falkowski PG (2010) The evolution and future of Earth's nitrogen cycle. Science 330(6001):192–196

Casaburi G, Duscher AA, Reid RP, Foster JS (2016) Characterization of the stromatolite microbiome from Little Darby Island, The Bahamas using predictive and whole shotgun metagenomic analysis. Environ Microbiol 18(5):1452–1469

Castro-Contreras SI, Gingras MK, Pecoits E, Aubet NR, Petrash D, Castro-Contreras SM, Dick G, Planavsky N, Konhauser KO (2014) Textural and geochemical features of freshwater. PALAIOS 29(5):192–209

Centeno CM, Legendre P, Beltran Y, Alcantara-Hernandez RJ, Lidstrom UE, Ashby MN, Falcón LI (2012) Microbialite genetic diversity and composition related to environmental variables. FEMS Microbiol Ecol 82:724–735

Chagas AA, Webb GE, Burne RV, Southam G (2016) Modern lacustrine microbialites: towards a synthesis of aqueous and carbonate geochemistry and mineralogy. Earth-Sci Rev 162:338–363

Coman C, Chiriac CM, Robeson MS, Ionescu C, Dragos N, Bardu-Tudoran L, Andrei AS, Banciu HL, Sicoria C, Podar M (2015) Structure, mineralogy, and microbial diversity of geothermal spring microbialites associated with a deep oil drilling in Romania. Front Microbiol 6:253

Couradeau E, Benzerara K, Moreira D, Gerard E, Kaźmierczak J, Tavera R, López-García P (2011) Prokaryotic and eukaryotic community structure in field and cultured microbialites from the alkaline Lake Alchichica (Mexico). PLoS One 6:e28767

Desnues C, Rodriguez-Brito B, Rayhawk S, Kelley S, Tran T, Haynes M, Liu H, Furlan M, Wegley L, Chau B, Ruan Y, Hall D, Angly FE, Edwards RA, Li L, Thurber RV, Reid RP, Siefert J, Souza V, Valentine DL, Swan BK, Breitbart M, Rohwer F (2008) Biodiversity and biogeography of phages in modern stromatolites and thrombolites. Nature 452(7185):340

Dupraz C, Reid RP, Braissant O, Dcho AW, Norman RS, Visscher PT (2009) Processes of carbonate precipitation in modern microbial mats. Earth Sci Rev 96(3):141–162

Dyhrman ST, Ammerman JW, Van Mooy BAS (2007) Microbes and the marine phosphorus cycle. Oceanography 20:110–116

Edgcomb VP, Bernhard JM, Summons RE, Orsi W, Beaudoin D, Visscher PT (2014) Active eukaryotes in microbialites from Highborne Cay, Bahamas, and Hamelin Pool (Shark Bay), Australia. ISME J 8:418

Falcón LI, Escobar-Briones E, Romero D (2002) Nitrogen fixation patterns displayed by cyanobacterial consortia in Alchichica crater-lake, Mexico. Hydrobiologia 467(1):71–78

Falkowski PG (1997) Evolution of the nitrogen cycle and its influence on the biological sequestration of CO_2 in the ocean. Nature 387(6630):272–275

Gamboa-Pérez HC, Schmitter-Soto JJ (1999) Distribution of cichlid fishes in the littoral of Lake Bacalar, Yucatan Peninsula. Environ Biol Fish 54(1):35–43

Gérard E, Ménez B, Couradeau E, Moreira D, Benzerara K, Tavera R, López-García P (2013) Specific carbonate–microbe interactions in the modern microbialites of Lake Alchichica (Mexico). ISME J 7(10):1997

Gischler E, Gibson MA, Oschmann W (2008) Giant Holocene freshwater microbialites, Laguna Bacalar, Quintana Roo, Mexico. Sedimentology 55:1293–1309

Gischler E, Golubic S, Gibson M, Oschamann W, Hudson JH (2011) Microbial mats and microbialites in the freshwater Laguna Bacalar, Yucatan Peninsula, Mexico. In: Reitner J, Sütwe T, Yuen D (eds) Advances in stromatolite geobiology. Springer, Berlin, pp 187–205

Gómez-Acata ES, Centeno CM, Falcón LI (2019) Methods for extracting'omes from microbialites. J Microbiol Methods 160:1–10

Hernández-Arana HA, Vega-Zepeda A, Ruíz-Zárate MA, Falcón-Alvarez LI, López-Adame H, Herrera-Silveira J, Kaster J (2015) Transverse coastal corridor: from freshwater lakes to coral reefs ecosystems. In: Islebe GA, Calmé S, León-Cortés JL, Schmook B (eds) Biodiversity and conservation of the Yucatán Peninsula. Springer, Cham, pp 355–376

Huang H, Shi P, Wang Y, Luo H, Shao N, Wang G, Yang P, Yao B (2009) Diversity of beta-propeller phytase genes in the intestinal contents of grass carp provides insight into the release of major phosphorus from phytate in nature. Appl Environ Microbiol 75(6):1508–1516

Johnson DB, Beddows PA, Flynn TM, Osburn MR (2018) Microbial diversity and biomarker analysis of modern freshwater microbialites from Laguna Bacalar, Mexico. Geobiology 16(3):319–337

Joye SB, Paerl HW (1994) Nitrogen cycling in microbial mats: rates and patterns of denitrification and nitrogen fixation. Mar Biol 119(2):285–295

Kaźmierczak J, Kempe S, Kremer B, López-García P, Moreira D, Tavera R (2011) Hydrochemistry and microbialites of the alkaline crater lake Alchichica, Mexico. Facies 57(4):543–570

Khodadad CLM, Foster JS (2012) Metagenomic and metabolic profiling of nonlithifying and lithifying stromatolitic mats of Highborne Cay, The Bahamas. PLoS One 7(5):e38229

Kolowith LC, Ingall ED, Benner R (2001) Composition and cycling of marine organic phosphorus. Limnol Oceanogr 46:309–320

Kruse CP, Basu P, Luesse DR, Wyatt SE (2017) Transcriptome and proteome responses in RNAlater preserved tissue of *Arabidopsis thaliana*. PLoS One 12:e0175943

Kurth D, Amadio A, Ordoñez OF, Albarracín VH, Gärtner W, Farías ME (2017) Arsenic metabolism in high altitude modern stromatolites revealed by metagenomic analysis. Sci Rep 7(1):1024

Lindsay MR, Anderson C, Fox N, Scofield G, Allen J, Anderson E, Bueter L, Poudel S, Sutherland K, Munson-McGee JH, Van Nostrand JD, Zhou J, Spear JR, Baxter BK, Lageson DR, Boyd ES (2017) Microbialite response to an anthropogenic salinity gradient in Great Salt Lake, Utah. Geobiology 15:131–145

López-García P, Kazmierczak J, Benzerara K, Kempe S, Guyton F, Moreira D (2005) Bacterial diversity and carbonate precipitation in the giant microbialites from the highly alkaline Lake Van, Turkey. Extremophiles 9:263–274

Louyakis AS, Gourlé H, Casaburi G, Bonjawo RM, Duscher AA, Foster JS (2018) A year in the life of a thrombolite: comparative metatranscriptomics reveals dynamic metabolic changes over diel and seasonal cycles. Environ Microbiol 20:842–861

Mancilla Villa OR, Bautista Olivas AL, Ortega Escobar HM, Sánchez Bernal EI, Can Chulim Á, Gutiérrez G, Manuel Y (2014) Hidrogeoquímica de salinas Zapotitlán y los lagos-cráter Alchichica y Atexcac, Puebla. Idesia (Arica) 32(1):55–69

Mobberley JM, Khodadad CLM, Visscher PT, Reid RP, Hagan P, Foster JS (2015) Inner workings of thrombolites: spatial gradients of metabolic activity as revealed by metatranscriptome profiling. Sci Rep 5:12601

Oren A (2011) Thermodynamic limits to microbial life at high salt concentrations. Environ Microbiol 13(8):1908–1923

Pepe-Ranney C, Berelson WM, Corsetti FA, Treants M, Spear JR (2012) Cyanobacterial construction of hot spring siliceous stromatolites in Yellowstone National Park. Environ Microbiol 14(5):1182–1197

Perry E, Velázquez-Oliman G, Marin L (2002) The hydrogeochemistry of the karst aquifer system of the northern Yucatan Peninsula, Mexico. Int Geol Rev 3:191–221

Perry E, Paytan A, Pedersen B, Velazquez-Oliman G (2009) Groundwater geochemistry of the Yucatan Peninsula, Mexico: constraints on stratigraphy and hydrogeology. J Hydrol 367(1–2):27–40

Ragon M, Benzerara K, Moreira D, Tavera, López-García P (2014) 16S rDNA-based analysis reveals cosmopolitan occurrence but limited diversity of two cyanobacterial lineages with contrasted patterns of intracellular carbonate mineralization. Front Microbiol 5

Riding R (2011) Microbialites, stromatolites, and thrombolites. Encycl Geobiol:635–654

Ruttenberg KC, Dyhrman ST (2005) Temporal and spatial variability of dissolved organic and inorganic phosphorus, and metrics of phosphorus bioavailability in an upwelling-dominated coastal system. J Geophys Res Oceans 110(C10)

Ruvindy R, White RA III, Neilan BA, Burns BP (2016) Unravelling core microbial metabolisms in the hypersaline microbial mats of Shark Bay using high-throughput metagenomics. ISME J 10(1):183–196

Saghaï A, Zivanovic Y, Zeyen N, Moreira D, Benzerara K, Deschamps P, Bertolino P, Ragon M, Tavera R, López-Archilla AI, López-García P (2015) Metagenome-based diversity analyses suggest a significant contribution of non-cyanobacterial lineages to carbonate precipitation in modern microbialites. Front Microbiol 6:797

Santos F, Peña A, Nogales B, Soria-Soria E, del Cura MÁG, González-Martín JA, Antón J (2010) Bacterial diversity in dry modern freshwater stromatolites from Ruidera pools Natural Park, Spain. Syst Appl Microbiol 33:209–221

Steppe TF, Pinckney JL, Dyble J, Paerl HW (2001) Diazotrophy in modern marine Bahamian stromatolites. Microb Ecol 41:36–44

Suzumura M, Kamatani A (1995) Origin and distribution of inositol hexaphosphate in estuarine and coastal sediments. Limnol Oceanogr 40:1254–1261

Tavera R, Komárek J (1996) Cyanoprokaryotes in the volcanic lake of Alchichica, Puebla State, Mexico. Arch Hydrobiol 117:511–538

Valdespino-Castillo PM (2015) Identification of the differential role of bacterial communities in the P cycle: microbialites and bacterioplankton from Alchichica lake as study models. PhD Dissertation. PCML, Universidad Nacional Autónoma de México, DF, Mexico

Valdespino-Castillo PM, Alcántara-Hernández RJ, Alcocer J, Merino-Ibarra M, Macek M, Falcón LI (2014) Alkaline phosphatases in microbialites and bacterioplankton from Alchichica soda lake, Mexico. FEMS Microbiol Ecol 90(2):504–519

Valdespino-Castillo PM, Alcántara-Hernández RJ, Merino-Ibarra M, Alcocer J, Macek M, Moreno-Guillén OA, Falcón LI (2017) Phylotype dynamics of bacterial P utilization genes in microbialites and bacterioplankton of a monomictic endorheic lake. Microb Ecol 73(2):296–309

Valdespino-Castillo PM, Hu P, Merino-Ibarra M, López-Gómez LM, Cerqueda-García D, González-De Zayas R, Pi-Puig T, Lestayo JA, Holman HY, Falcón LI (2018) Exploring biogeochemistry and microbial diversity of extant microbialites in Mexico and Cuba. Front Microbiol 9:510

Velázquez NIT, Vieyra MR, Paytan A, Broach KH, Terrones LMH (2018) Hydrochemistry and carbonate sediment characterization of Bacalar Lagoon, Mexican Caribbean. Mar Freshw Res 70(3):382–394

Warden JG, Casaburi G, Omelon CR, Bennett PC, Breecker DO, Foster JS (2016) Characterization of microbial mat microbiomes in the modern thrombolite ecosystem of Lake Clifton, Western Australia using shotgun metagenomics. Front Microbiol 7:1064

White RW III, Allen R, Wong HL, Ruvindy R, Neilan B, Burns BP (2018) Viral communities of Shark Bay modern stromatolites. Front Microbiol 9:1223

Wong HL, Smith D-L, Visscher PT, Burns BP (2015) Niche differentiation of bacterial communities at a millimeter scale in Shark Bay microbial mats. Sci Rep 5:15607

Yanez-Montalvo A, Gómez-Acata S, Águila B, Hernández-Arana H, Falcón LI, Senko JM (2020) The microbiome of modern microbialites in Bacalar Lagoon, Mexico. PLoS ONE 15 (3):e0230071

Zhu T, Dittrich M (2016) Carbonate precipitation through microbial activities in natural environment, and their potential in biotechnology: a review. Front Bioeng Biotechnol 4:4

Chapter 10
The Origin and Early Evolution of Life on Earth: A Laboratory in the School of Science

José Alberto Campillo-Balderas and Arturo Becerra

Abstract The study and research in the origin of life in Mexico have deep roots in the last century. However, the modern period of study of the origin and early evolution of life started in the early 1970s, when two groups were created to study the topic at the Universidad Nacional Autónoma de México (UNAM). The first of these groups began in 1974 in the School of Sciences and has continued to work since then: analyzing comets and meteors as models of chemical events that occurred in the early Earth; studying the central hypothesis of the "RNA world" and the role of modified ribonucleotides on the early stages of evolution; using bioinformatics tools and comparative genomics; and trying to infer very early stages of life. Here, we present a brief description of the main lines of research of the first group and the perspectives of the study of the origin and early evolution of life. This chapter addresses a question that does not have a model site, since the environmental conditions where it arose are still unknown. Nevertheless, many inferences can be made out of experiments and comparative genomics.

10.1 The Study of the Origin and Early Evolution of Life in Mexico

Study and research of the origin of life in Mexico have deep roots in the last century since Alfonso L. Herrera, a remarkable naturalist and teacher, introduced the evolutionary theory in basic education programs and proposed his theory of plasmogeny, as a physicochemical process of the origin of life (Perezgasga et al. 2003; Ledesma-Mateos and Barahona 2003). Herrera made an essential improvement in Mexican education and life sciences. As Negrón-Mendoza said, "He is considered a corner-

J. A. Campillo-Balderas · A. Becerra (✉)
Facultad de Ciencias, Universidad Nacional Autónoma de México, UNAM,
Mexico City, Mexico
e-mail: campillo@ciencias.unam.mx; abb@ciencias.unam.mx

© Springer Nature Switzerland AG 2020
V. Souza et al. (eds.), *Astrobiology and Cuatro Ciénegas Basin as an Analog of Early Earth*, Cuatro Ciénegas Basin: An Endangered Hyperdiverse Oasis,
https://doi.org/10.1007/978-3-030-46087-7_10

stone of Biology in Mexico" (Negrón-Mendoza 1995). When Herrera had published his hypotheses (Herrera 1904, 1905, 1919, 1924), Alexander I. Oparin was close to announcing his foundational views about the heterotrophic origin of life, whose conclusions were that living organisms are the historical result of a gradual transformation of lifeless matter (Lazcano 2016). First, in a book (Oparin 1924), which had no international notice, since only a few Russian scientists read it, but then in 1938, an English edition of Oparin's book was published and became known worldwide (Oparin 1938), where Oparin attempted to provide in detail a chemical process for the origin of life (Miller et al. 1997).

In our view, the modern period of study of early evolution of life in Mexico could have been started in the early 1970s, around the moment that Oparin visited our university (Universidad Nacional Autónoma de México, UNAM), thanks to the academic authorities of the university and the invitation of the young scientist Antonio Lazcano. This event was not only an excellent opportunity to listen to the most crucial scientist on the field, but also it was a critical moment that allowed the foundation of the modern view of the study of origins of life in our university and Mexico. A few years later, were created in our university at least two new groups working in the origin of life: one by Antonio Lazcano at the School of Sciences in 1974 and the other by the chemist Alicia Negrón at the Institute of Nuclear Science, the laboratory of chemical evolution in 1976. In conjunction with research, Professor Lazcano founded the first course about "Origin of Life" that has been imparted from 1974 until now as a requested and successful program on the School of Science. Both laboratories were and are thriving and vital groups that generate not only several and outstanding research but also a good number of scientists who trained as students in both groups (Chap. 1). In this chapter, we write about the main activities and investigations of the first one.

10.2 The Origin of Life Laboratory in the School of Science

10.2.1 Prebiotic Chemistry and RNA World

In the beginning, the laboratory in the School of Science at the UNAM was called *Laboratorio A. I. Oparin*; it was founded in September 1974 by Antonio Lazcano. Since this time, he had been working on the importance of extraterrestrial organic compounds present in nuclei of comets and chondritic meteorites, next to Joan Oró, Stanley L. Miller, and other prominent pioneers of the field. They analyzed comets and meteors as models of chemical events that occurred in the solar system when the Earth itself was forming, to explain the synthesis of organic molecules that could provide the conditions of the primitive Earth (Oró et al. 1978, 1980; Bar-Nun et al. 1981; Lazcano et al. 1983). Also, significant meetings occurred in the field on those years including the second and third Mexican meetings of Origin of Life (Fig. 10.1); the first as a tribute to Oparin in 1975; and the "Cosmochemistry and the Origins of Life" where Joan Oró, Cyril Ponnamperuma, Lynn Margulis, George Fox, and other founders of the field presented their results and research.

Fig. 10.1 The third Mexican meetings of Origin of Life. "Cosmochemistry and the Origins of Life" where Joan Oró, Cyril Ponnamperuma, Lynn Margulis, George Fox presented their results and research in 1979

The discovery of the catalytic activities of RNA molecules motivated Lazcano to propose in 1986 the existence of what we now call the "RNA world," a central hypothesis in the modern study of the emergence of life, and which was suggested the same year, independently, by Walter Gilbert and Bruce Alberts (Lazcano 1986). Since that year until today, the group of "Origin of Life Laboratory" addressed the study of this hypothesis, which has opened a line of research that seeks to understand the origin of modified ribonucleotides (Fig. 10.2), such as coenzymes, alarmones, and histidine (Hernández-Morales et al. 2019; Vázquez-Salazar et al. 2018). Nevertheless, how life originated on Earth is not fully knowable, the development of catalytic RNAs and the diverse roles that RNA molecules and ribonucleotides play in extant cells (Vázquez-Salazar and Lazcano 2018) has provided substantial support to the possibility that this genetic polymer carried out a much more conspicuous role on the early stages of evolution, an epoch generally referred to as the RNA world (Gilbert 1986).

Some of the main results in this subject and obtained by the founder, students, and colleagues of the laboratory are (a) the demonstration of the catalytic activities of pure peptides under prebiotic conditions (Shen et al. 1990); (b) the display that it is possible to synthesize organic compounds of biochemical importance in reducing and neutral atmospheres, which broke a paradigm that for the past 60 years indicated otherwise (Bada et al. 2007; Cleaves et al. 2008); (c) the demonstration of the similarity of formation of alkylated amino acids in chondritic meteorites and neutral

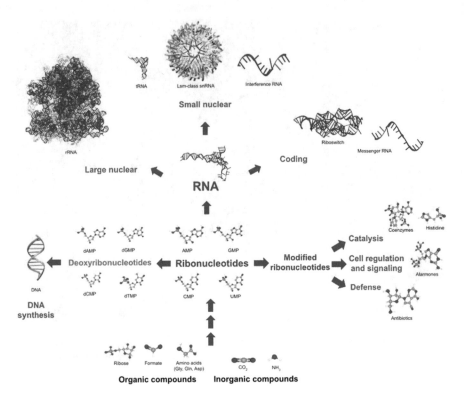

Fig. 10.2 The role of ribonucleotides, such as coenzymes, alarmones, and histidine, in the origin and early evolution of life

and reducing atmospheres (Parker et al. 2011a, b); and (d) the analysis of the original samples of the Miller experiment, which led to a research article that was published in the journal *Science* (Johnson et al. 2008; Bada and Lazcano 2003).

10.2.2 Early Life and the Last Common Ancestor

Despite the unbeatable obstacles surrounding the understanding of early life, a key question in astrobiology, there are two ways to approach it, especially to the evolutionary transition between the non-living and the living, to the process that we call the origin of life: (1) the synthetic approach or also called "bottom-up approach" and (2) the reconstruction of early entities using the actual organisms or the "top-down approach" (Fig. 10.3).

Our group has been using both approaches, mainly since the genomic era started when the first two genomes were available and the comparison of them was plausible. The first completely sequenced genome of a living organism was *Haemophilus influenzae* (Fleischmann et al. 1995) then *Mycoplasma genitalium* a few months

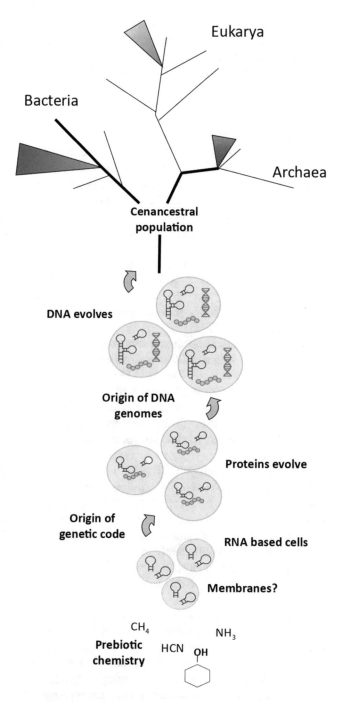

Fig. 10.3 Approaches to early life. The synthetic approach or also called "bottom-up approach" and the reconstruction of early entities using the actual organisms or the "top-down approach"

later (Fraser et al. 1995). At that moment we started to compare the genomes to infer the gene complement and metabolic abilities of early stages, continuing a lineage of research that began years before (Lazcano et al. 1988, 1992).

However, Mushegian and Koonin (1996) published the first inference of the comparison of both genomes. They concluded that the information obtained suggested a minimal gene set for cellular life and reflected some characteristics of the last common ancestor of life, a cell that was a straightforward one that lacked a DNA genome and complex metabolic pathways. Nevertheless, we propose that polyphyletic gene losses can be bias the characterizations of the cenancestor (Becerra et al. 1997). In our proposal, we insist that in the characterization of ancestral states, should be considered the biological aspects as well as the bioinformatic or statistical data since we are doomed to have partial reconstructions and are not exempt from false-negative results. Furthermore, a bioinformatics approach to the origin of life is not feasible since all possible intermediates that may have once existed have long since being lost.

Phylogenetic analyses based on comparative genomics provide essential clues on the very early stages of biological evolution and diversification of life. Still, it is hard to see how its applicability can be used beyond a time that corresponds to a cellular evolution period in which protein biosynthesis was already working (Becerra et al. 2007; Lazcano 2012).

In other words, phylogenetic analysis and comparative genomics cannot give us direct information about the origin of life. Sensitive tools can help us infer the very early stages of cellular evolution, as Woese and Fox discovered that, regarding the similarities and differences in the small subunit of the ribosomal RNA molecule, living beings are divided into three groups (Woese and Fox 1977a). They suggested that there was a primitive entity in the division of these three lineages when the relation between genotype and phenotype had not yet evolved (Woese and Fox 1977b; Delaye and Becerra 2012).

They named this ancestral entity a progenote, which implies a primitive state. However, thanks to the development of DNA sequencing technology and the accumulation of a vast amount and diversity of sequences and complete genomes in databases, it was possible to start identifying the genes conserved in bacteria and Archaea. Our group has a particular interest and has been working significantly in this area. We have collaborated importantly in the reconstruction and characterization of the cenancestor, also known as the last common ancestor (LCA), which means the last common ancestor of currently living beings (Delaye et al. 2005; Becerra and Delaye 2015).

Despite qualitative and quantitative differences in the methodology used to identify the gene complement of LCA, there is an overlap that reveals an impressive level of conservation for a significant number of sequences involved in essential biological processes. The results obtained include lists of repertoires of gene sequences from incompletely represented basic biological processes, such as translation, transcription, biosynthesis of nucleotides and amino acids, energy metabolism, sequences related to replication, and cellular transport (Muñoz-Velasco et al. 2018). Corroborating the proposal in 1992 by Lazcano, Fox, and Oró, that proposed that the last common ancestor possessed genetic machinery similar to that of a modern bacterium (Lazcano et al. 1992; Delaye and Becerra 2012).

Nevertheless, the current descriptions of LCA are limited by the scant information available, and the gene complement includes proteins that originated in different epochs prior to LCA time. For instance, several highly conserved ribosomal proteins may have originated during the RNA/protein world stage, whereas thymidine kinase and thioredoxin reductase, which are involved in deoxyribonucleotide biosynthesis, evolved at a later stage. Although we favor a bacterial-like ancestor, at this moment, it is difficult to assess the metabolic pathways related to the LCA or the environmental conditions in which it prospered (Becerra et al. 2007). Other research groups propose that acetogenesis and methanogenesis are the oldest metabolisms on Earth (Martin and Russell 2007; Sousa et al. 2013) and suggest that both metabolic routes emerged in a hyperthermophilic environment, such as the alkaline hydrothermal environment (Martin and Russell 2007), from an LCA endowed with geochemically driven monocarbon-unit transformations (Sousa and Martin 2014; Weiss et al. 2016). However, our results suggest that methanogenesis is present only in the Archaea domain, whereas the acetyl-CoA synthesis from CO_2, or Wood-Ljungdahl pathway, is present in both bacteria and Archaea (Muñoz-Velasco et al. 2018), but also we are interested in the metabolic adaptation to the oxygen that increased its level in the primitive atmosphere (Alvarez-Carreño et al. 2016).

10.2.3 Early Metabolic Evolution

Some authors suggest that the contemporary metabolic routes are descended from prebiotic chemical pathways and differ by the intervention of enzymes (Degani and Halmann 1967; Hartman 1975; de Duve 1991; Morowitz 1992). However, the known prebiotic pathways are quite different from the present metabolic routes, as Lazcano and Miller said: "the origin of metabolic pathways lies closer to the origin of life than to the last common ancestor" (Lazcano and Miller 1999). Therefore, to study the evolution and origin of the contemporary metabolic pathways, it gives limited or no knowledge about the process that occurred near the beginning of life.

However, it is also essential to understand and inquire what kind of metabolism was present in early life (close to LCA) and infer how was the process to assemble the pathways and increase their number of enzymes. There are four main hypotheses about the origin and assembly of metabolic pathways:

(a) The Horowitz hypothesis was the first attempt to explain this evolutionary process and was proposed by Horowitz in 1945. It is also referred to as the retrograde hypothesis, which means the pathways were built up backward a step at a time than in the forward direction using intermediates in the prebiotic conditions (Horowitz 1945).

(b) The Granick hypothesis is the development of the biosynthetic pathway that was assembled in the forward direction, where the prebiotic compounds do not perform a role (Granick 1950, 1957).

(c) The Patchwork assembly hypothesis, an exciting idea proposed independently by Ycas in 1974 and Jensen in 1976, according to which biosynthetic pathways are the outcome of the serial recruitment of promiscuous enzymes endowed with broad catalytic specificity that could react in different steps and routes.

(d) The semi-enzymatic origin of metabolic pathways, proposed by Lazcano and Miller in 1999, suggests that the metabolic-like pathways produced crucial components required by primitive living entities, and were originally non-enzymatic or semi-enzymatic autocatalytic processes. That later became fine-tuned by ribozyme- and protein-based metabolic routes (Lazcano and Miller 1999; Islas et al. 1998; Delaye and Lazcano 2005).

Unlike the origin of life, the early evolution of metabolic pathways has directly benefited from the development of genomics and molecular evolution. Thanks to the comparative analysis of complete sequences of prokaryotic and eukaryotic genomes, it is possible to study the origin and evolution of metabolism. For instance, the hypotheses mentioned above relate to a greater or lesser degree with gene duplication, which is a crucial mechanism of genome evolution that affects genomic complexity, intrinsic genetic diversity (Innan and Kondrashov 2010), and is ultimately a significant source of evolutionary innovation, as Lewis (1951) and Ohno (1970) proposed it for the development of metabolic changes (Fani 2012). Our group works on the subject since the bioinformatic tools are available, analyzing the enzymatic substrate specificity and its implication in the early evolution of metabolic pathways (Lazcano et al. 1994); the role of gene duplication in the evolution of purine nucleotide salvage pathways (Becerra and Lazcano 1998); the evolution of histidine biosynthesis (Fani et al. 1995, 1997, 1998; Alifano et al. 1996); the evolution of autotrophy metabolism (Peretó et al. 1999; Becerra et al. 2014); and the evolution of methanogenesis pathways (Muñoz-Velasco et al. 2018).

10.2.4 The Origin of Life and Viruses

In the first half of the twentieth century, there was a dichotomic explanation to the origin of life that persists until now. For one side, the metabolism-first hypothesis is based on the idea that life is "a dynamic, self-regulating, metabolizing system." Biochemists, like Oparin, promoted it. On the other side, the nucleocentric hypothesis emphasizes that life is "a single molecule with the inherent capacity to undergo such essential activities as self-duplication." Geneticists, like Hermann Muller, were the ones who supported the idea of an "autocatalytic enzyme/gene" as a primordial form of life (Podolsky 1996).

Precisely, in 1926, Muller proposed that viruses, as small primitive genes, self-replicating, non-metabolizing agents, are the relics of a primordial form of life.

Even Jerome Alexander, Calvin Bridges, and, later, J.B.S. Haldane (then retracted) reinforced this idea because they believed that viruses were the missing link between nonlife and the primordial cells (Podolsky 1996; López-García 2012). With the discovery of DNA in 1944 and its description as a molecule with genetic information and evolutionary characteristics, the metabolism-first hypothesis lost strength. Therefore, viruses seen as living fossils of a primitive living entity, the nucleocentric became the virocentric hypothesis of the origin of life and viruses (Moreira and López-García 2009) (Fig. 10.4a).

Later in the 1930s, Robert Green and Patrick Laidlaw (Green 1935; Laidlaw 1939) challenged the virocentric with their retrograde hypothesis of the origin of viruses (Fig. 10.4b). According to them, all viruses descended from a cellular ancestor that degenerated and subsequently increased the level of parasitism. In 1945, Burnet recognized the debate between those two hypotheses (virocentric and degenerate), but also he empirically proposed that viruses could be escaped fragments from modern cells (Burnet 1945) (Fig. 10.4c). This hypothesis was also supported later in 1970 with the discovery of the reverse transcriptase and the proposal of RNA tumor viruses as mobile genetic elements of animal cells (Baltimore 1970; Mizutani and Temin 1970; Temin 1970).

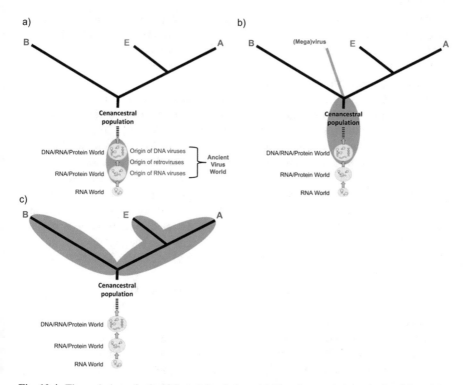

Fig. 10.4 The main hypothesis of the origin of virus. (**a**) The virocentric hypothesis of the origin of life and viruses. (**b**) The retrograde hypothesis of the origin of viruses. (**c**) The escape hypothesis

Virus research in the latter half of the twentieth century has allowed an increase in the knowledge of the evolution of viruses. However, their origin is still a debate because viruses are polyphyletic. With the sequencing of new viral genomes, the free access to viral databases, and overall, with the expertise in the origin and early evolution of life, our laboratory wanted to be part of the origin-of-viruses debate in 2014. We realized that there were new refined versions of the three hypotheses. With the findings of supposedly hallmark genes like RNA-dependent RNA polymerase (replication) and capsid (structure) and the nature and size of viral genomes, it has been proposed that RNA viruses originated in the prebiotic RNA world: a kind of an "ancient virus world" or a virosphere that preceded the contemporary biosphere (Koonin et al. 2006; Forterre 2006) (Fig. 10.4a). Later, with the discovery of some double-stranded DNA megaviruses in 2001 and the analyses of phylogenetic markers involved in DNA replication and repair, transcription, and translation, viruses have been seen as a consequence of a degenerate origin of an ancestral cell and even proposed as a fourth domain of life (Boyer et al. 2010; Claverie et al. 2006; Nasir et al. 2015) (Fig. 10.4b). Finally, with the studies of horizontal gene transfer and blocks of genes as functional modules, viruses could have a chimeric-escaped gene origin (Campbell 2001; Hendrix et al. 2000) (Fig. 10.4c).

We are aware of the highly divergent nature of viral genomes, which has compromised the elucidation of deep phylogenetic relationships (Holmes 2009). Therefore, we have based our work on the origin and early evolution of viruses in three strategies: the study of all biological and ecological data, pangenomic analysis, and protein structure similarities for each viral family.

In a comparative data analysis of the chemical nature, genome size, and segmentation, and host type of all viral references reported in GenBank, we have shown that the viral genome size, like RNA viruses, does not correlate with the evolutionary history of hosts (Campillo-Balderas et al. 2015). We found that RNA viral families are well distributed in eukaryotic hosts. With the possibility of bias, *Cystoviridae* and *Leviviridae* are the only two RNA viral families which infect prokaryotes like *Pseudomonas*, Caulobacter, *Acinetobacter*, and pathogenic Enterobacteriaceae. All of these hosts are found in the microbiome of many animals. There are no RNA viruses which infect Archaea. Accordingly, it appears that RNA viruses have a recent origin closely related to the evolutionary history of eukaryotes. It can thus be concluded that RNA viruses may be ancient, but not primitive (Fig. 10.5).

Once we have designed the database with all the biological information, we have been analyzing the pangenome of all families of DNA and RNA viruses at this moment. We recruited the reference sequences of all proteins from GenBank, searched for homologs, and clustered them into protein families and pangenomic compartments (core, dispensable, and unique proteins) depending on their presence and absence in each viral family. With this approach (Contreras-Moreira and Vinuesa 2013), we expect to trace the biological history of all proteins that are shared by each viral family and to determine whether they stem from cells or viruses according to their distribution in phylogenies.

Fig. 10.5 RNA viral families distributed in the tree of life. RNA viruses only infect the Eukarya domain. There are no RNA viruses in Archaea and only two families that infect *Proteobacteria* (*Pseudomonas*, Caulobacter, *Acinetobacter*, and pathogenic Enterobacteriaceae). This bacterial group coexist with eukaryotes; therefore, these viruses probably evolved from the eukaryotic viruses

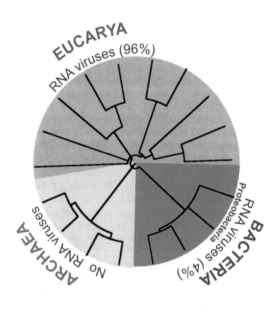

We also want to analyze some tertiary structures like capsid proteins. A recent study based on structural similarities has shown that capsid proteins probably evolved from ancestral proteins of cells on multiple occasions (Krupovic and Koonin 2017). In our laboratory, we have designed an approach for inferring phylogenies based on the construction of matrices of pairwise structural comparison scores (Jácome et al. 2015). We expect to determine a deep phylogenetic relationship of jelly-roll domain, the most prevalent fold of icosahedral capsid proteins, based on this methodology.

With these preliminary results, we want to determine a consilience and discordance of findings according to the biological and ecological information, the pangenomic analysis, and deep phylogenies of tertiary structures of all viral families. With this, we expect to elaborate an origin and early evolution model, supporting the idea that viruses are escaping genes from host genomes.

Finally, the physicochemical conditions in which life started cannot be determined unequivocally, and how life originated on Earth is unknowable. Nevertheless, the development of cosmology, geology, and prebiotic chemistry allows more robust inferences about this scenario. This synthetic approach together with the top-down approach releases not only the reconstruction of patterns of early stages of life but also an opportunity to learn and discover the process of evolution. Using prebiotic chemistry, bioinformatics tools, comparative genomics, but above all, evolutionary biology, students and professors from the Origin of Life laboratory in the School of Science, we try to investigate one of the most exciting stages of life, the very early stages of evolution.

Acknowledgments The authors would like to thank all the members of the Origen de la Vida Laboratory at the Universidad Nacional Autónoma de México. We also acknowledge the help of Dr. Ricardo Hernández-Morales and Dr. Alberto Vázquez-Salazar.

References

Alifano P, Fani R, Liò P, Lazcano A, Bazzicalupo M, Carlomagno MS, Bruni CB (1996) Histidine biosynthetic pathway and genes: structure, regulation, and evolution. Microbiol Rev 60:44–69

Alvarez-Carreño C, Becerra A, Lazcano A (2016) Molecular evolution of the oxygen-binding hemerythrin domain. PLoS One 11(6):e0157904

Bada JL, Lazcano A (2003) Prebiotic soup: revisiting the Miller experiment. Science 300:745–746

Bada JL, Fegley B, Miller SL, Lazcano A, Cleaves HJ, Hazen RM, Chalmers J (2007) Debating evidence for the origin of life. Science 315:937–938

Baltimore D (1970) Viral RNA-dependent DNA polymerase: RNA-dependent DNA polymerase in virions of RNA tumour viruses. Nature 226:1211

Bar-Nun A, Lazcano-Araujo A, Oró J (1981) Could life have evolved in cometary nuclei? Orig Life Evol Biosph 11:387–394

Becerra A, Delaye L (2015) The universal ancestor. Mètode 87:1

Becerra A, Lazcano A (1998) The role of gene duplication in the evolution of purine nucleotide salvage pathways. Orig Life Evol Biosph 28:539–553

Becerra A, Islas S, Leguina JI, Silva E, Lazcano A (1997) Polyphyletic gene losses can bias backtrack characterizations of the cenancestor. J Mol Evol 45:115–117

Becerra A, Delaye L, Islas A, Lazcano A (2007) Very early stages of biological evolution related to the nature of the last common ancestor of the three major cell domains. Annu Rev Ecol Evol Syst 38:361–379

Becerra A, Rivas M, García-Ferris C, Lazcano A, Peretó J (2014) A phylogenetic approach to the early evolution of autotrophy: the case of the reverse TCA and the reductive acetyl-CoA pathways. Int Microbiol 17(2):91–97

Boyer M, Madoui MA, Gimenez G, La Scola B, Raoult D (2010) Phylogenetic and phyletic studies of informational genes in genomes highlight existence of a 4 Domain of life including giant viruses. PLoS One 5(12):e15530

Burnet Sir FM (1945) Virus as organism: evolutionary and ecological aspects of some human virus diseases. Harvard University Press, Cambridge

Campbell A (2001) The origins and evolution of viruses. Trends Microbiol 9(2):61

Campillo-Balderas JA, Lazcano A, Becerra A (2015) Viral genome size distribution does not correlate with the antiquity of the host lineages. Front Ecol Evol 3:143

Claverie JM, Ogata H, Audic S, Abergel C, Suhre K, Fournier PE (2006) Mimivirus and the emerging concept of 'giant' virus. Virus Res 117(1):133–144

Cleaves JH, Chalmers JH, Lazcano A, Miller SL, Bada JL (2008) Prebiotic organic synthesis in neutral planetary atmospheres. Orig Life Evol Biosph 38:105–155

Contreras-Moreira B, Vinuesa P (2013) GET_HOMOLOGUES, a versatile software package for scalable and robust microbial pangenome analysis. Appl Environ Microbiol 79(24):7696–7701

de Duve C (1991) Blueprint for a cell. Neil Patterson, Burlington

Degani C, Halmann M (1967) Chemical evolution of carbohydrate metabolism. Nature 216:1207

Delaye L, Becerra A (2012) Cenancestor, the last universal common ancestor. Evol Educ Outreach 5:382–388

Delaye L, Lazcano A (2005) Prebiological evolution and the physics of the origin of life. Phys Life Rev 2(1):47–64

Delaye L, Becerra A, Lazcano A (2005) The last common ancestor: what's in a name? Orig Life Evol Biosph 35:537–554

Fani R (2012) The origin and evolution of metabolic pathways: why and how did primordial cells construct metabolic routes? Evol Educ Outreach 5:367–381

Fani R, Liò P, Lazcano A (1995) Molecular evolution of the histidine biosynthetic pathway. J Mol Evol 41:760–774

Fani R, Barberio C, Casalone E, Cavalieri D, Lazcano A, Liò P, Mori E, Perito B, Polsinelli M (1997) Paralogous histidine biosynthetic genes: evolutionary analysis of the *Saccharomyces cerevisiae* HIS6 and HIS7 genes. Gene 197:9–17

Fani R, Mori E, Tamburini E, Lazcano A (1998) Evolution of the structure and chromosomal distribution of histidine biosynthetic genes. Orig Life Evol Biosph 28:555–570

Fleischmann RD, Adams MD, White O, Clayton RA, Kirkness EF, Kerlavage AR, Bult CJ, Tomb JF, Dougherty BA, Merrick JM et al (1995) Whole-genome random sequencing and assembly of *Haemophilus influenzae* Rd. Science 269(5223):496–512

Forterre P (2006) The origin of viruses and their possible roles in major evolutionary transitions. Virus Res 117(1):5–16

Fraser CM, Gocayne JD, White O, Adams MD, Clayton RA, Fleischmann RD, Bult CJ, Kerlavage AR, Sutton G, Kelley JM, Fritchman RD, Weidman JF, Small KV, Sandusky M, Fuhrmann J, Nguyen D, Utterback TR, Saudek DM, Phillips CA, Merrick JM, Tomb JF, Dougherty BA, Bott KF, Hu PC, Lucier TS, Peterson SN, Smith HO, Hutchison CA, Venter JC (1995) The minimal gene complement of *Mycoplasma genitalium*. Science 270(5235):397–403

Gilbert W (1986) Origin of life: the RNA world. Nature 319:618

Granick S (1950) The structural and functional relationships between heme and chlorophyll. Harvey Lect 44:220–224

Granick S (1957) Speculations on the origins and evolution of photosynthesis. Ann N Y Acad Sci 69:292–230

Green R (1935) On the nature of the filterable viruses. Science 82:444

Hartman H (1975) Speculations on the origin and evolution of metabolism. J Mol Evol 4:359–370

Hendrix RW, Lawrence JG, Hatfull GF, Casjens S (2000) The origins and ongoing evolution of viruses. Trends Microbiol 8(11):504–508

Hernández-Morales R, Becerra A, Lazcano A (2019) Alarmones as vestiges of a bygone RNA world. J Mol Evol 87(1):37–51

Herrera AL (1904) Nociones de Biología, Primera Edición, Secretaría de Fomento

Herrera AL (1905) Una Nueva Ciencia: La Plasmogenesis. Bol. de Instrucción Pública, México, pp 606–625

Herrera AL (1919) Some studies in plasmogenesis. J Lab Clin Med 4(8):479–483

Herrera AL (1924) Biología y Plasmogenia, Herrero Hermanos Sucesores, 3rd edn, México

Holmes EC (2009) The evolution and emergence of RNA viruses. Oxford University Press, New York

Horowitz NH (1945) On the evolution of biochemical syntheses. Proc Natl Acad Sci U S A 31:153–157

Innan H, Kondrashov F (2010) The evolution of gene duplications: classifying and distinguishing between models. Nat Rev Genet 11(2):97–10

Islas S, Becerra A, Leguina JI, Lazcano A (1998) Early metabolic evolution: insights from comparative genomics. In: Chela Flores J, Raulin F (eds) Trieste conference on chemical evolution. V. Exobiology: matter, energy, and information in the origin and evolution of life in the universe. Kluwer Academic Publishers, Dordrecht, pp 167–174

Jácome R, Becerra A, Ponce de León S, Lazcano A (2015) Structural analysis of monomeric RNA-dependent polymerases: evolutionary and therapeutic implications. PLoS One 10(9):e0139001

Jensen RA (1976) Enzyme recruitment in evolution of new function. Annu Rev Microbiol 30:409–425

Johnson AP, Cleaves HJ, Dworkin JP, Glavin DP, Lazcano A, Bada JL (2008) The Miller volcanic spark discharge experiment. Science 322:404

Koonin EV, Senkevich TG, Dolja VV (2006) The ancient virus world and evolution of cells. Biol Dir 1:29

Krupovic M, Koonin EV (2017) Multiple origins of viral capsid proteins from cellular ancestors. Proc Natl Acad Sci U S A 114(12):E2401–E2410

Laidlaw P (1939) Virus diseases and viruses. Cambridge University Press, Cambridge, p 32

Lazcano A (1986) Prebiotic evolution and the origin of cells. Treballs Soc Catalana Biol 39:73–103

Lazcano A (2012) The origin and early evolution of life: where, when and how? Evol Educ Outreach 5:334–336

Lazcano A (2016) Alexander I. Oparin and the origin of life: a historical reassessment of the heterotrophic theory. J Mol Evol 83:214

Lazcano A, Miller S (1999) On the origin of metabolic pathways. J Mol Evol 49:424–431

Lazcano A, Oró J, Miller SL (1983) Primitive Earth environments: organic synthesis and the origin and early evolution of life. Precambrian Res 20:259–282

Lazcano A, Guerrero R, Margulis L, Oró J (1988) The evolutionary transition from RNA to DNA in early cells. J Mol Evol 27:283–290

Lazcano A, Fox GE, Oró J (1992) Life before DNA: the origin and early evolution of early Archean cells. In: Mortlock RP (ed) The evolution of metabolic function. CRC Press, Boca Raton, pp 237–295

Lazcano A, Díaz-Villagómez E, Mills T, Oró J (1994) On the levels of enzymatic substrate specificity: implications for the early evolution of metabolic pathways. Adv Space Res 15:345–356

Ledesma-Mateos I, Barahona A (2003) The institutionalization of biology in Mexico in the early 20th century. The conflict between Alfonso Luis Herrera (1868–1942) and Isaac Ochoterena (1885–1950). J Hist Biol 36(2):285–307

Lewis EB (1951) Pseudoallelism and gene evolution. Cold Spring Harb Symp Quant Biol 16:159–174

López-García P (2012) The place of viruses in biology in light of the metabolism- versus-replication-first debate. Hist Philos Life Sci 34(3):391–406

Martin W, Russell MJ (2007) On the origin of biochemistry at an alkaline hydrothermal vent. Philos Trans R Soc Lond Ser B Biol Sci 362(1486):1887–1925

Miller S, Schopf J, Lazcano A (1997) Oparin's "origin of life": sixty years later. J Mol Evol 44:351–353

Mizutani S, Temin HM (1970) An RNA-dependent DNA polymerase in virions of rous sarcoma virus. Cold Spring Harb Symp Quant Biol 35:847–849

Moreira D, López-García P (2009) Ten reasons to exclude viruses from the tree of life. Nat Rev Microbiol 7(4):306–311

Morowitz HJ (1992) Beginnings of cellular life: metabolism recapitulates biogenesis. Yale University Press, New Haven

Muñoz-Velasco I, García-Ferris C, Hernandez-Morales R, Lazcano A, Peretó J, Becerra A (2018) Methanogenesis on early stages of life: ancient but not primordial. Orig Life Evol Biosph 48(4):407–420

Mushegian AR, Koonin EV (1996) A minimal gene set for cellular life derived by comparison of complete bacterial genomes. Proc Natl Acad Sci U S A 93:10268–10273

Nasir A, Sun FJ, Kim KM, Caetano-Anollés G (2015) Untangling the origin of viruses and their impact on cellular evolution. Ann N Y Acad Sci 1341:61–74

Negrón-Mendoza A (1995) Alfonso L. Herrera: a Mexican pioneer in the study of chemical evolution. J Biol Phys 20:11

Ohno S (1970) Evolution by gene duplication. Springer, Berlin

Oparin AI (1924) Proiskhozhdenie zhizny. Izd. Moskovski Rabochii, Moscow

Oparin AI (1938) The origin of life. Macmillan, New York

Oró J, Holzer G, Lazcano-Araujo A (1978) Organic cosmochemistry and the origins of life. In: Oró J (ed) Proceedings of the workshop on experimental aspects of comets. Houston, Lunar and Planetary Institute, pp 54–59

Oró J, Holzer G, Lazcano-Araujo A (1980) The contribution of cometary volatiles to the primitive Earth. Life Sci Space Res 18:67–82

Parker ET, Cleaves JH, Dworkin JP, Glavin DP, Callahan MP, Aubrey AD, Lazcano A, Bada JL (2011a) Primordial synthesis of amines and amino acids in a 1958 Miller H2S-rich spark discharge experiment. Proc Natl Acad Sci U S A 108:5526–5531

Parker ET, Cleaves JH, Callahan MP, Dworkin JP, Glavin DP, Lazcano A, Bada JL (2011b) Enhanced synthesis of alkyl amino acids in Miller's 1958 H2S experiment. Orig Life Evol Biosph 41:569–574

Peretó J, Velasco AM, Becerra A, Lazcano A (1999) Comparative biochemistry of CO2 fixation and the evolution of autotrophy. Int Microbiol 2:3–10

Perezgasga L, Silva E, Lazcano A, Negrón-Mendoza A (2003) The sulfocyanic theory on the origin of life: towards a critical reappraisal of an autotrophic theory. Int J Astrobiol 2:301

Podolsky S (1996) The role of the virus in origin-of-life theorizing. J Hist Biol 29(1):79–126

Shen C, Lazcano A, Oró J (1990) Enhancement activities of histidyl-histidine in some prebiotic reactions. J Mol Evol 31:445–452

Sousa FL, Martin WF (2014) Biochemical fossils of the ancient transition from geoenergetics to bioenergetics in prokaryotic one carbon compound metabolism. BBA-Bioenergetics 1837(7):964–981

Sousa FL, Thiergart T, Landan G, Nelson-Sathi S, Pereira I, Allen JF, Lane N, Martin WF (2013) Early bioenergetic evolution. Philos Trans R Soc Lond Ser B Biol Sci 368(1622):20130088

Temin HM (1970) Malignant transformation of cells by viruses. Perspect Biol Med 14(1):11–26

Vázquez-Salazar A, Lazcano A (2018) Early life: embracing the RNA world. Curr Biol 28(5):220–222

Vázquez-Salazar A, Becerra A, Lazcano A (2018) Evolutionary convergence in the biosyntheses of the imidazole moieties of histidine and purines. PLoS One 13(4):e0196349

Weiss MC, Sousa FL, Mrnjavac N, Neukirchen S, Roettger M, Nelson-Sathi S, Martin WF (2016) The physiology and habitat of the last universal common ancestor. Nat Microbiol 1(9):1–8

Woese CR, Fox GE (1977a) Phylogenetic structure of the prokaryotic domain: the primary kingdoms. Proc Natl Acad Sci U S A 74:5088–5090

Woese CR, Fox GE (1977b) The concept of cellular evolution. J Mol Evol 10:1–6

Ycas M (1974) On earlier states of the biochemical system. J Theor Biol 44:145–160

Chapter 11
Cuatro Ciénegas as an Archaean Astrobiology Park

Nahui Olin Medina-Chávez, Susana De la Torre-Zavala,
Alejandra E. Arreola-Triana, and Valeria Souza

"Exobiology is no more fantastic than the realization of space travel itself, and we have a grave responsibility to explore its implications for science and for human welfare with our best scientific insights and knowledge."

Joshua Lederberg 1960, Science

Abstract The Cuatro Ciénegas Basin (CCB), located in Coahuila, México, is considered a living laboratory for astrobiology. This site is an analogue for the Gale Crater in Mars because of its gypsum-rich soils. Moreover, the "pozas" at the site are extremely oligotrophic, which is characteristic of ancient oceans chemistry. Despite the oligotrophic conditions, CCB harbors a very rich microbial diversity in the form of stromatolites and microbial mats. Archaea data at CCB has been sparse given that, in general, archaea do not represent more than 1% of the reads in a metagenome. Here, we describe a new hyper-diverse site with an abundant core of Archaea. This site, which we named "archaean domes," recreates in their interior the atmosphere of ancient Earth and Mars and can serve as a valuable resource for astrobiological research.

N. O. Medina-Chávez · S. De la Torre-Zavala
Instituto de Biotecnología, Universidad Autónoma de Nuevo León, Facultad de Ciencias Biológicas, San Nicolás de los Garza, NL, Mexico

A. E. Arreola-Triana
Departamento de Biología Celular y Genética, Universidad Autónoma de Nuevo León, Facultad de Ciencias Biológicas, San Nicolás de los Garza, NL, Mexico

V. Souza (✉)
Instituto de Ecología, Universidad Nacional Autónoma de México, UNAM, Mexico City, Mexico
e-mail: souza@unam.mx

© Springer Nature Switzerland AG 2020
V. Souza et al. (eds.), *Astrobiology and Cuatro Ciénegas Basin as an Analog of Early Earth*, Cuatro Ciénegas Basin: An Endangered Hyperdiverse Oasis,
https://doi.org/10.1007/978-3-030-46087-7_11

11.1 Cuatro Ciénegas Basin: A Living Laboratory for Astrobiology

In 1976, the Viking landers performed experiments looking for life on Mars. These experiments were designed to search for Class I organisms—those that live in relatively moderate temperatures (> -33 °C) and high water activity ($a_w > 0.7$), like most of the organisms on Earth. The experiments yielded negative, but controversial, results. Some of the controversy arose from the fact that these experiments were not designed to look for organisms belonging to Classes II to IV, that is, organisms that live in extreme conditions with low temperatures and high water activity, high temperature and low water activity, or low temperature and low water activity (Sagan and Lederberg 1976). More than 40 years ago, Sagan and Lederberg suggested that future space crafts that landed on Mars should be equipped to detect the full range of organisms that could potentially inhabit the planet. To more fully understand the putative Martian life and the constraints it lives under, it is essential to use analogue ecosystems on Earth.

Analogues mimic and hypothesize the origin and occurrence of life beyond Earth. Astrobiologists on Earth work in analogue environments that are similar to those found in space. Of particular interest are the extreme environments where life could not be thought possible (Gupta et al. 2014; Jones et al. 2018). Soare et al. (2001) describe three types of analogues. Analogues of the first order are based on empirical evidence gathered from spacecraft, landers, and satellite images; on Earth, features, such as impact craters, volcanoes, and aeolian dunes, serve as analogues of the first order. Analogues of the second order are those based on indirect but highly suggestive evidence, such as alluvial fans and gypsum veins, the latter of which Steve Squyres called "the single most powerful evidence for liquid water on Mars" (Szynkiewicz et al. 2010; Showstack 2011). Analogues of the third order are speculative, unsupported by direct or indirect evidence—as long as there is no confirmation of life existing elsewhere in the solar system, extremophile organisms are analogues of the third order (Preston and Dartnell 2014).

The study of extremophiles has educated our search for life in other planets. Of the terrestrial bodies in the Solar System, Mars is the most Earth-like, and features of its geology and geography point to a warmer, wetter past, as evidenced by the traces of gypsum identified by the OMEGA instrument on ESA's Mars Express orbiter (Bibring et al. 2005; Paige 2005; Preston and Dartnell 2014). On Earth, there are over 30 sites that emulate the extreme conditions found on Mars, among which are the Atacama Desert (Azua-Bustos et al. 2012), the Ka'u Desert (Seelos et al. 2010), the Mojave Desert (Salas et al. 2011), Rio Tinto (López-Archilla et al. 2001; Amils et al. 2014), McMurdo Dry Valley (Chan-Yam et al. 2019), and permafrost soils (Douglas and Mellon 2019). In Mexico there are two Mars analogues: the salt flats of Guerrero Negro, Baja California Sur, and the Cuatro Ciénegas Basin, Coahuila.

The Cuatro Ciénegas Basin (CCB) is an arid, gypsum-rich environment that is considered an analogue of the Gale Crater and Olympia Undae on Mars (Szynkiewicz et al. 2010; López-Lozano et al. 2012). The CCB is located at the bottom of a valley in the northern Mexican state of Coahuila, at an altitude of 740 masl (Souza et al. 2012). The area receives approximately 200 mm of precipitation, and the ponds, or "pozas," are fed through extremely oligotrophic groundwater aquifers whose stoichiometry resembles that of the early Earth. More than 200 pozas in the CCB are considered hyper-diverse because of their heterogeneity and endemism found in the microbial communities.

The geological description has been addressed previously (Meyer 1973; Escalante et al. 2008; Souza et al. 2012, 2018a; Hipkin et al. 2013), but the main picture exposes the CCB as an extreme scenery, due to its inner arid conditions that allow high levels of oligotrophy, especially for phosphorus—the CCB reportedly has the lowest levels of this element. Also, the high rate of radiation and the salty emergence due to mineral and ion precipitation confers the CCB features that are similar to the Martian landscape (López-Lozano et al. 2012; Souza et al. 2012).

Despite the oligotrophic state of the water in the CCB and the stoichiometric levels of phosphorous that are below those needed to maintain DNA and protein synthesis, the CCB has a remarkable level of biodiversity in all domains and taxa and is one of the few sites in the world with living stromatolites (Elser et al. 2005; Souza et al. 2006, 2008, 2012, 2018a; Alcaraz et al. 2008).

The research of these living fossil formations and the sediment rearrangements (microbial mats) can give us a glimpse of how life could have looked on the early stages this planet and others. Recently, CCB was described as a new evolutionary model called a "lost world" (Souza et al. 2018b).

Multidisciplinary research teams are working at this place to demonstrate our hypothesis: that a seed bank of ancestral lineages has been trapped inside the deep aquifer of a mountain with a magmatic pouch under its sediments; when water from the aquifer rises to the surface and refills the pozas, it brings with it the deep biosphere from the seed bank. This mountain, named San Marcos y Pinos, has been giving shelter and isolation to these relictual communities for thousands or millions of years and allowing speciation and divergence to take place (Souza et al. 2018b).

A particular hydrological system within CCB, the Churince, possesses a large diversity of microorganisms from the Bacteria and Archaea domains, specifically 5167 operational taxonomic units (OTUs) with a 97% identity from 16s rRNA gene sequencing (Souza et al. 2018b). These identifications suggest that CCB has a marine ancestry with certain phyla like Actinobacteria, Bacteroidetes, and Proteobacteria (Souza et al. 2018b).

These extreme physicochemical and environmental conditions, along with the ancient microbial communities found at this site, make the CCB a model for the Earth's earliest ecosystems—a Precambrian Park—and a perfect analogue for the gypsum-rich Martian soils (López-Lozano et al. 2012; Souza et al. 2012).

CCB shows that life can not only hide and survive under the sediments but also transform these extreme environments through their metabolisms. We hypothesize that the study of the CCB can gleam information about the ancient Martian sea that left its imprint in the sediments of the central mount of the Gale Crater and help missions, such as the Mars Science Laboratory, trace the biosignatures left by changes in chemical stoichiometry through the Martian eras. If life exists or has existed in the Gale Crater, it is very possible that it left a biosignature imprint in the gypsum crystals (Edwards et al. 2006). In this case, a firm understanding of an analogous region such as the CCB would be invaluable to the search for life on Mars (Newsom et al. 2001; Szynkiewicz et al. 2010; López-Lozano et al. 2012).

11.2 Stromatolites and Microbial Mats in CCB: Revelations of the History of Life

Stromatolites are the legacy of early life on Earth, representing a treasure trove for studies of paleoecology and evolution. These formations give an insight into microbial interactions, their role in the appearance of Eukaryotes, and how changes in the atmosphere and climate have affected this and other planets with the "ideal" conditions for life (McNamara and Awramik 1992; Breitbart et al. 2009). More information about stromatolites can be found in Chap. 4 of this book.

The formation of stromatolites has decreased since the appearance of plants and metazoans (Awramik 1971), and living stromatolites can be found in a few localities including Hamelin Pool, Shark Bay, Australia (Reid et al. 2003; Burns et al. 2004; Papineau et al. 2005), Yellowstone National Park, United States (Berelson et al. 2011), and in the Cuatro Ciénegas Basin (CCB) (García-Oliva et al. 2018; Souza et al. 2018b). The bacterial communities in the CCB stromatolites rely on the availability of light, water depth, current flow, and desiccation (Paterson et al. 2008).

A vast number of studies have proved that stromatolites and microbial mats in the CCB have a "marine signature," implying that the whole ecosystem may be considered relics, coming from an ancient marine community (Souza et al. 2006; Desnues et al. 2008).

In 2008, Desnues et al. (2008) performed a comparative metagenomic analysis of viral communities from the freshwater stromatolites in Pozas Azules II (PAII) and Rio Mesquites (RM) within the CCB and the marine stromatolites from Highborne Cay, Bahamas. They compared their sequences with nonredundant Genbank and SEED databases and found that the viral communities in the stromatolites were unrelated, with 98.8% unique sequences for Highborne Cay, 99.3% for PAII, and 97.7% for Rio Mesquites. Assembly of contigs shows that freshwater stromatolites (PAII and RM) did not produce any cross-contigs with marine microbialites (Highborne Cay), suggesting that none of the viruses was shared between them (Desnues et al. 2008).

One year later, Breitbart et al. (2009) conducted metagenomics and isotopic analyses of the stromatolites in PAII and RM. They found that both samples had Bacteria dominance with an 87% in PAII and 95% in RM samples. Interestingly, a comparison against SEED database distinguished 1061 (PAII) and 203 (RM) archaeal sequences, and more than 86% belonged to Euryarchaeota phylum. Their findings also showed that although the stromatolites at these sites are different in function and taxa, the genes involved in ammonia assimilation, nitrogen fixation, assimilatory nitrate reduction, and phosphorus metabolism genes were present in both metagenomes, demonstrating a great adaptation to the CCB thanks to polyphosphate accumulations that serve a reservoir. It has been proposed that stromatolites are present in the CCB because of its stoichiometric nutrient constraints (Elser et al. 2005), which cause the coordinated metabolic activity from the whole microbial community to create microenvironments that favor carbonate precipitation (Breitbart et al. 2009).

Besides stromatolites, CCB also harbors microbial mats, another extraordinary well-organized biological structure. A microbial mat is a vertically stratified microbial community, defined by physicochemical gradients that support most biogeochemical cycles in aquatic environments, making these structures self-sufficient (Bolhuis et al. 2004). In 2012, Peimbert and collaborators (2012) did a metagenomic comparison of two aquatic microbial mats (red and green mats). They found that the red mat lacked phosphorus and was dominated with *Pseudomonas*, whereas the green mat was nitrogen-limited with *Cyanothece* as the most abundant genus, making these mats metabolically different and diverse, although heterotrophic taxa was shared between the two mats.

For the reasons described above, the stromatolites and microbial mats in the CCB are perfect astrobiology analogues of the third order. These successful and efficient microbial communities are capable of cycling the few nutrients available to them, just like their counterparts did during the Archean Eon.

11.3 The Archaean Domes in Cuatro Ciénegas Basin: A Lost World Enclosing a Secret of the Tough Archaean Life

The exploration of the CCB in March 2016 led by Dr. Valeria Souza identified "Archaean Domes" (AD) located near the Poza Azul II, within the Pozas Azules Ranch. Archaean Domes are a very particular site that under dry conditions, which span most of the year, looks just like another salty place. However, after a heavy rain, the water dissolves the salty crust allowing not only photosynthesis to be less constricted but, more importantly, establishing a communication between the deep biosphere and the surface. Under wet conditions, a complex system of dome-like structures forms in this unique, shallow, clay-rich pond. Inside, these domes are dominated by methane-rich gases that simulate the conditions of the late Archean Eon (Medina-Chávez et al. 2019), whereas the outer shell presents all the photosynthetic lineages as well as several fungi and protozoa (Espinosa-Asuar, to be submitted). During the dry season in the CCB, the AD area dries up, leaving a crystalline salty crust on the top layer of the mat (Medina-Chávez et al. 2019).

Microbial activity can be limited by desiccation stress and low substrate diffusion (Schimel et al. 2007; Aanderud et al. 2015), causing dormancy events; microbial metabolism is reactivated once the soils are soaked again (Fierer and Schimel

2003; Aanderud et al. 2015). These dormancy processes driven by harsh conditions could be key players in these biosphere fluctuations, leading to microbial community maintenance. Similar conditions of stress may be happening in other planets, and dormancy periods like the ones described could make detection of life more difficult.

Results revealed that AD mats comprise an abundant, hyper-diverse, and cohesive archaeal core community that maintains its abundance and diversity through time and despite environmental fluctuations. Members of five phyla (Euryarchaeota, Crenarchaeota, and the unexpected Thaumarchaeota, Korarchaeota, and Nanoarchaeota) displayed 230 species (OTU's at 97% cutoff) corresponding to a 5% of archaeal relative abundance from a whole-genome SGS analysis (Medina-Chávez et al. 2019), which is an uncommon value even for other hypersaline microbial mats (Fernandez et al. 2016; Babilonia et al. 2018). It is worth noting that a study of genetic diversity of archaea among stromatolites in several regions in Mexico, including the CCB, showed a low abundance (up to 1.76% of the community) of Archaea (Centeno et al. 2012). Similar results were obtained from the analysis of two aquatic microbial mats from the CCB ponds in 2012, where Archaea abundance reached only 2% of the community in a green mat (Peimbert et al. 2012). Taking into account that the Archaean Domes have not been extensively explored and sampled as other archaeal-rich sites (Wong et al. 2016), the AD in CCB undoubtedly represent a model to study how archaeal communities thrive in conditions similar to the expected to be found on exoplanets.

The AD site in CCB is extremely saline, reaching up to 53% during the wet season and staying under saturation the rest of the year. The site is surrounded by a carbonate layer and poor vegetation. Because the microbial mats can be found only beneath the salt and not the carbonate, it is likely that salinity is an important abiotic factor for the community below. In the AD, fluctuations in the relative abundances of rare taxa through time seem to be related to precipitation; the behavior of the archaeal community is consistent with the findings of a study on community networks in a (non-hypersaline) microbial mat within the Churince Pond in the CCB, which demonstrates that halotolerant taxa increases during dry conditions (De Anda et al. 2018). Chloride and sulfates are characteristic minerals on Mars, suggesting that brines and other hypersaline environments may have been a part of Martian history (Martínez and Renno 2013; Smith et al. 2014). If this were the case, then halophilic microorganisms could be part of the "rare taxa" (Lynch and Neufeld 2015) in the biota of another planet.

NASA's Curiosity rover found different levels and concentrations of methane throughout its mission. In June 2019, it detected the highest levels of methane (Greicius 2019), a relevant finding considering that there are only two possible sources for the production of this gas: the interaction given by rocks and water (in this case, frozen water) and microbial communities (Sherwood Lollar et al. 2007; Etiope and Sherwood Lollar 2013; Dean et al. 2018). If the methane found on Mars is biogenically produced, it is likely to be the product of organisms similar to Euryarchaeota, which we found in high abundance throughout the seasons in the hypersaline, desiccated AD.

Polyextremophilia is another phenomenon expected and studied for Mars analogues. Halophilia and alkalophilia are often found together, for example, Yellowstone and Rio Tinto are both hypersaline and alkaline (White et al. 2015). If there are microbes on Mars, then we can hypothesize that Archaea or other Bacteria of an extreme nature would be the most likely candidates. These domains contain methanogenic bacteria and polyextremophiles that are adapted to survive not one but several extreme conditions at once (Mesbah and Wiegel 2008; Seckbach 2013). Salinity is not the only abiotic factor leading the presence of extremophilic microorganisms—temperature, pressure, oxygen availability, and pH also play a major role on the composition and structure of microbial communities. The AD of CCB, besides its extreme salinity, reaches pH values of 10 during wet seasons (data not shown), whereas the pH at the Poza Azul II at a distance of 50 m from the AD reaches 7.9 (Centeno et al. 2012; Espinosa-Asuar et al. 2015).

11.4   Conclusions

Ever since its discovery, the Cuatro Ciénegas Basin has not ceased to amaze us. The site harbors a richness of endemic macro and microbiota, and its cohesive microbial community thrives in extreme environments, building niches that seem relict from an ancient sea and adapting to the feedback and biological input of the Anthropocene.

The oligotrophic, gypsum-rich, and poly-extreme environments in CCB have recently been described as an invaluable ecological and functional model for the study of adaptations and evolution in harsh, fluctuating environments. This puts the domain Archaea, for the first time, as a core component of the CCB microbiome, with a fundamental role in the maintenance of the hyper-diversity of this particular site. The study of the ecology and the evolution of the Archaean Domes in the CCB can take us deeper into a "lost world" that will continue to help us understand the cycles and biosignatures that life, past or present, may leave on Martian or other habitable extraterrestrial soils.

Acknowledgments We want to thank Dr. Laura Espinosa-Asuar and Erika Aguirre Planter for their technical help. To Pronatura Noreste for their help in logistics and getting access to their ranch and APFF Cuatrociénegas for their support .This research was funded by PAPIIT- DGAPA, UNAM grant IG200319 to VS and Luis E. Eguiarte.

References

Aanderud ZT, Jones SE, Fierer N, Lennon JT (2015) Resuscitation of the rare biosphere contributes to pulses of ecosystem activity. Front Microbiol 6:24
Alcaraz LD, Olmedo G, Bonilla G, Cerritos R, Hernández G, Cruz A, Ramírez E, Putonti C, Jiménez B, Martínez E, López V (2008) The genome of *Bacillus coahuilensis* reveals

adaptations essential for survival in the relic of an ancient marine environment. Proc Natl Acad Sci U S A 105(15):5803–5808

Amils R, Fernández-Remolar D, the IPBSL Team TI (2014) Río Tinto: a geochemical and mineralogical terrestrial analogue of Mars. Life 4:511–534

Awramik SM (1971) Precambrian columnar stromatolite diversity: reflection of metazoan appearance. Science 174:825–827

Azua-Bustos A, Urrejola C, Vicuña R (2012) Life at the dry edge: microorganisms of the Atacama Desert. FEBS Lett 586:2939–2294

Babilonia J, Conesa A, Casaburi G, Pereira C, Louyakis AS, Reid RP, Foster JS (2018) Comparative metagenomics provides insight into the ecosystem functioning of the Shark Bay stromatolites, Western Australia. Front Microbiol https://doi.org/10.3389/fmicb.2018.01359

Berelson WM, Corsetti FA, Pepe-Ranney C, Hammond DE, Beaumont W, Spear JR (2011) Hot spring siliceous stromatolites from Yellowstone National Park: assessing growth rate and laminae formation. Geobiology 9(5):411–424

Bibring JP, Langevin Y, Gendrin A, Gondet B, Poulet F, Berthé M, Soufflot A, Arvidson R, Mangold N, Mustard J, Drossart P (2005) Mars surface diversity as revealed by the OMEGA/Mars Express observations. Science 307:1576–1581

Bolhuis H, Poele EM, Rodriguez-Valera F (2004) Isolation and cultivation of Walsby's square archaeon. Environ Microbiol 6:1287–1291

Breitbart M, Hoare A, Nitti A, Siefert J, Haynes M, Dinsdale E, Edwards R, Souza V, Rohwer F, Hollander D (2009) Metagenomic and stable isotopic analyses of modern freshwater microbialites in Cuatro Ciénegas, Mexico. Environ Microbiol 11:16–34

Burns BP, Goh F, Allen M, Neilan BA (2004) Microbial diversity of extant stromatolites in the hypersaline marine environment of Shark Bay, Australia. Environ Microbiol 6:1096–1101

Centeno CM, Legendre P, Beltrán Y, Alcántara-Hernández RJ, Lidström UE, Ashby MN, Falcón LI (2012) Microbialite genetic diversity and composition relate to environmental variables. FEMS Microbiol Ecol 82:724–735

Chan-Yam K, Goordial J, Greer C, Davila A, McKay CP, Whyte LG (2019) Microbial activity and habitability of an Antarctic Dry Valley water track. Astrobiology 19:757–770

De Anda V, Zapata-Peñasco I, Blaz J, Poot-Hernandez AC, Contreras-Moreira B, Hernandez Rosales M, Eguiarte LE, Souza V (2018) Understanding the mechanisms behind the response to environmental perturbation in microbial mats: a metagenomic-network based approach. Front Microbiol 9:2606

Dean JF, Middelburg JJ, Röckmann T, Aerts R, Blauw LG, Egger M, Jetten MS, de Jong AE, Meisel OH, Rasigraf O, Slomp CP (2018) Methane feedbacks to the global climate system in a warmer world. Rev Geophys 56:207–250

Desnues C, Rodriguez-Brito B, Rayhawk S, Kelley S, Tran T, Haynes M, Liu H, Furlan M, Wegley L, Chau B, Ruan Y (2008) Biodiversity and biogeography of phages in modern stromatolites and thrombolites. Nature 452(7185):340–343

Douglas TA, Mellon MT (2019) Sublimation of terrestrial permafrost and the implications for ice-loss processes on Mars. Nat Commun 10:1716

Edwards HG, Mohsin MA, Sadooni FN, Hassan NF, Munshi T (2006) Life in the sabkha: Raman spectroscopy of halotrophic extremophiles of relevance to planetary exploration. Anal Bioanal Chem 385(1):46–56

Elser JJ, Schampel JH, Garcia-Pichl FE, Wade BD, Souza V, Eguiarte L, Escalante AN, Farmer JD (2005) Effects of phosphorus enrichment and grazing snails on modern stromatolitic microbial communities. Freshw Biol 50:1808–1825

Escalante AE, Eguiarte LE, Espinosa-Asuar L, Forney LJ, Noguez AM, Souza SV (2008) Diversity of aquatic prokaryotic communities in the Cuatro Ciénegas basin. FEMS Microbiol Ecol 65:50–60

Espinosa-Asuar L, Escalante AE, Gasca-Pineda J, Blaz J, Peña L, Eguiarte LE, Souza V (2015) Aquatic bacterial assemblage structure in Pozas Azules, Cuatro Ciénegas Basin, Mexico: deterministic vs. stochastic processes. Int Microbiol 18:105–115

Etiope G, Sherwood Lollar B (2013) Abiotic methane on Earth. Rev Geophys 51:276–299

Fernandez AB, Rasuk MC, Visscher PT, Contreras M, Novoa F, Poire DG, Patterson MM, Ventosa A, Farias ME (2016) Microbial diversity in sediment ecosystems (Evaporites Domes, Microbial Mats and Crusts) of Hypersaline Laguna Tebenquiche, Salar de Atacama, Chile. Front Microbiol 7:1284. https://doi.org/10.3389/fmicb.2016.01284

Fierer N, Schimel JP (2003) A proposed mechanism for the pulse in carbon dioxide production commonly observed following the rapid rewetting of a dry soil. Soil Sci Soc Am J 67:798–805

García-Oliva F, Tapia-Torres Y, Montiel-Gonzalez C, Perroni-Ventura Y (2018) Carbon, nitrogen, and phosphorus in terrestrial pools: where are the main nutrients located in the grasslands of the Cuatro Ciénegas Basin? In: García-Oliva F, Elser J, Souza V (eds) Ecosystem Ecology and Geochemistry of Cuatro Cienegas. How to Survive in an Extremely Oligotrophic Site. Springer, Cham, pp 1–13

Greicius T (2019) Curiosity's Mars methane mystery continues. https://www.nasa.gov/feature/jpl/curiosity-detects-unusually-high-methane-levels. Accessed 26 Sept 2019

Gupta GN, Srivastava S, Khare SK, Prakash V (2014) Extremophiles: an overview of microorganism from extreme environment. Int J Agric Environ Biotechnol 7:371

Hipkin VJ, Voytek MA, Meyer MA, Léveillé R, Domagal-Goldman SD (2013) Analogue sites for Mars missions: NASA's Mars science laboratory and beyond – overview of an international workshop held at the Woodlands, Texas, on March 5–6, 2011. Icarus 224:261–267

Jones RM, Goordial JM, Orcutt BN (2018) Low energy subsurface environments as extraterrestrial analogs. Front Microbiol 9:1605

Lederberg J (1960) Exobiology: approaches to life beyond the earth. Science 132 (3424):393–400. https://doi.org/10.1126/science.132.3424.393

López-Archilla AI, Marin I, Amils R (2001) Microbial community composition and ecology of an acidic aquatic environment: the Tinto River, Spain. Microb Ecol 41:20–35

López-Lozano NE, Eguiarte LE, Bonilla-Rosso G, García-Oliva F, Martínez-Piedragil C, Rooks C, Souza V (2012) Bacterial communities and the nitrogen cycle in the gypsum soils of Cuatro Ciénegas Basin, Coahuila: a Mars analogue. Astrobiology 12:699–709

Lynch MDJ, Neufeld JD (2015) Ecology and exploration of the rare biosphere. Nat Rev Microbiol 13:217–229

Martínez GM, Renno NO (2013) Water and brines on Mars: current evidence and implications for MSL. Space Sci Rev 175:29–51

McNamara KJ, Awramik SM (1992) Stromatolites: a key to understanding the early evolution of life. Sci Prog 76:345–364

Medina-Chávez NO, Viladomat-Jasso M, Olmedo-Álvarez G, Eguiarte LE, Souza V, De la Torre-Zavala S (2019) Diversity of archaea domain in Cuatro Cienegas Basin: Archean Domes. Submitted, 1–39

Mesbah NM, Wiegel J (2008) Life at extreme limits. Ann N Y Acad Sci 1125:44–57

Meyer ER (1973) Late-Quaternary paleoecology of the Cuatro Cienegas Basin, Coahuila, Mexico. Ecology 54:982–995

Newsom HE, Hagerty JJ, Thorsos IE (2001) Location and sampling of aqueous and hydrothermal deposits in Martian impact craters. Astrobiology 1:71–88

Paige DA (2005) Ancient Mars: wet in many places. Science 307:1575–1576

Papineau D, Walker JJ, Mojzsis SJ, Pace NR (2005) Composition and structure of microbial communities from stromatolites of Hamelin Pool in Shark Bay, Western Australia. Appl Environ Microbiol 71:4822–4832

Paterson DM, Aspden RJ, Visscher PT, Consalvey M, Andres MS, Decho AW, Stolz J, Reid RP (2008) Light-dependant biostabilisation of sediments by stromatolite assemblages. PLoS One 3:e3176

Peimbert M, Alcaraz LD, Bonilla-Rosso G, Olmedo-Alvarez G, García-Oliva F, Segovia L, Eguiarte LE, Souza V (2012) Comparative metagenomics of two microbial mats at Cuatro Ciénegas Basin I: ancient lessons on how to cope with an environment under severe nutrient stress. Astrobiology 12(7):648–658

Preston LJ, Dartnell LR (2014) Planetary habitability: lessons learned from terrestrial analogues. Int J Astrobiol 13:81–98

Reid RP, James NP, Macintyre IG, Dupraz CP, Burne RV (2003) Shark Bay stromatolites: microfabrics and reinterpretation of origins. Facies 49(1):299

Sagan C, Lederberg J (1976) The prospect of life on Mars: a pre-Viking assessment. Icarus 28(2):291–300

Salas EC, Abbey W, Bhartia R, Beegle LW (2011) The Mojave desert: a Martian analog site for future astrobiology themed missions.42 Lunar and Planetary Institute Conference, Woodlands, Texas, March 7–11, 2011, http://hdl.handle.net/2014/43479

Schimel J, Balser TC, Wallenstein M (2007) Microbial stress-response physiology and its implications for ecosystem function. Ecology 88:1386–1394

Seckbach J (2013) Life on the edge and astrobiology: who is who in the polyextremophiles world? In: Polyextremophiles. Springer, Dordrecht, pp 61–79

Seelos KD, Arvidson RE, Jolliff BL, Chemtob SM, Morris RV, Ming DW, Swayze GA (2010) Silica in a Mars analog environment: Ka'u Desert, Kilauea Volcano, Hawaii. J Geophys Res 115(E4)

Sherwood Lollar B, Voglesonger K, Lin LH, Lacrampe-Couloume G, Telling J, Abrajano TA, Onstott TC, Pratt LM (2007) Hydrogeologic controls on episodic H_2 release from Precambrian fractured rocks—energy for deep subsurface life on Earth and Mars. Astrobiology 7:971–986

Showstack R (2011) Mars Opportunity rover finds gypsum veins. Eos Trans AGU 92(51):479

Smith ML, Claire MW, Catling DC, Zahnle KJ (2014) The formation of sulfate, nitrate and perchlorate salts in the Martian atmosphere. Icarus 231:51–64

Soare R, Pollard W, Green D (2001) Deductive model proposed for evaluating terrestrial analogues. Eos Trans AGU 82(43):501

Souza V, Espinosa-Asuar L, Escalante AE, Eguiarte LE, Farmer J, Forney L, Lloret L, Rodríguez-Martínez JM, Soberón X, Dirzo R, Elser JJ (2006) An endangered oasis of aquatic microbial biodiversity in the Chihuahuan desert. Proc Natl Acad Sci U S A 103:6565–6570

Souza V, Eguiarte LE, Siefert J, Elser JJ (2008) Microbial endemism: does phosphorus limitation enhance speciation? Nat Rev Microbiol 6:559–564

Souza V, Siefert JL, Escalante AE, Elser JJ, Eguiarte LE (2012) The Cuatro Ciénegas basin in Coahuila, Mexico: an astrobiological Precambrian park. Astrobiology 12:641–647

Souza V, Eguiarte LE, Elser JJ, Travisano M, Olmedo-Álvarez G (2018a) A microbial Saga: how to study an unexpected hot spot of microbial biodiversity from scratch? In: Souza V, Olmedo-Álvarez G, Eguiarte LE (eds) Cuatro Ciénegas ecology, natural history and microbiology. Springer, Cham, pp 1–20

Souza V, Moreno-Letelier A, Travisano M, Alcaraz LD, Olmedo G, Eguiarte LE (2018b) The lost world of Cuatro Ciénegas Basin, a relictual bacterial niche in a desert oasis. eLife 7:e38278

Szynkiewicz A, Ewing RC, Moore CH, Glamoclija M, Bustos D, Pratt LM (2010) Origin of terrestrial gypsum dunes-implications for Martian gypsum-rich dunes of Olympia Undae. Geomorphology 121:69–83

White RA III, Power IM, Dipple GM, Southam G, Suttle CA (2015) Metagenomic analysis reveals that modern microbialites and polar microbial mats have similar taxonomic and functional potential. Front Microbiol 6:966

Wong H, Ahmed-Cox A, Burns B (2016) Molecular ecology of hypersaline microbial mats: current insights and new directions. Microorganisms 4(1):6

Index

© Springer Nature Switzerland AG 2020
V. Souza et al. (eds.), *Astrobiology and Cuatro Ciénegas Basin as an Analog of
Early Earth*, Cuatro Ciénegas Basin: An Endangered Hyperdiverse Oasis,
https://doi.org/10.1007/978-3-030-46087-7

Printed in the United States
by Baker & Taylor Publisher Services